Marketing Scientific and Technical Information

Other Titles of Interest

Technological Innovation: The Experimental R&D Incentives Program, edited by Donald E. Cunningham, John R. Craig, and Theodore Schlie

Federal Funding of Civilian Research and Development, edited by Michael Michaelis

The State of Science and Research: Some New Indicators, edited by Nestor E. Terleckyj

Science, Politics, and Controversy: Civilian Nuclear Power in the United States, 1946-1974, Steven L. Del Sesto

Managing Earth From Space, Arthur Levine

Westview Special Studies in Information Management

Marketing Scientific and Technical Information
edited by William R. King and Gerald Zaltman

Creating and disseminating scientific and technical information (STI) can be likened to producing and distributing a product or service. Although this view is natural to marketing scholars and practitioners, it is not one that has been extensively applied to STI policymaking and research. This book assesses and demonstrates the applicability and potential of various areas of marketing theory in the STI context. It includes the work of distinguished marketing scholars who have analyzed STI marketing from such perspectives as consumer needs assessment, information acquisition strategy, market segmentation, and product design.

William R. King is professor of business administration at the University of Pittsburgh. Gerald Zaltman is Albert Wesley Frey Professor of Marketing in the Graduate School of Business, University of Pittsburgh.

Marketing Scientific and Technical Information

edited by William R. King and Gerald Zaltman

Routledge
Taylor & Francis Group

LONDON AND NEW YORK

First published 1979 by Westview Press

Published 2018 by Routledge
52 Vanderbilt Avenue, New York, NY 10017
2 Park Square, Milton Park, Abingdon, Oxon OX14 4RN

Routledge is an imprint of the Taylor & Francis Group, an informa business

Library of Congress Cataloging in Publication Data
Main entry under title:
Marketing scientific and technical information.
    (Westview Special Studies in Information Management)
    Papers presented at a conference sponsored by the National Science Foundation and held at the Graduate School of Business, University of Pittsburgh.
    Includes bibliographical references.
    1. Technical literature–Marketing. 2. Scientific literature–Marketing. I. King, William Richard, 1938-   II. Zaltman, Gerald. III. United States. National Science Foundation.
T10.7.M37   658.8'09'507   79-13350
ISBN 0-89158-397-1

ISBN 13: 978-0-367-02092-7 (hbk)
ISBN 13: 978-0-367-17079-0 (pbk)

Contents

Tables and Figures

Preface

This book is a compilation of papers that were presented at a conference on the marketing of scientific and technical information held at the Graduate School of Business of the University of Pittsburgh. The conference, sponsored by the National Science Foundation, brought together marketing scholars and practitioners in the dissemination of scientific and technical information.

The basic hypothesis of the conference was that marketing theory, as developed and applied in consumer and industrial product and service contexts, might be applied to the marketing of scientific and technical information. The objective of the conference was the preliminary translation of areas of marketing theory into forms that might be appropriate for the STI context and the preliminary evaluation of these theory areas in terms of their applicability and potential in STI dissemination.

A secondary objective of the conference was that of developing a professional interchange between marketing theorists and STI practitioners. Such an exchange of views should have the effect of familiarizing practitioners with the potentially useful marketing theory that is available for their use and of familiarizing marketing professionals with the state-of-the-art and with opportunities for research that may exist in the STI environment.

The conference was preceded by a process of investigations conducted by the editors. This process involved a preliminary evaluation of the areas of marketing theory that are most likely to be productive in the STI context and the problems and opportunity areas in STI dissemination. These evaluations were made on the basis of a review of the STI literature, discussions with marketing theorists, and a set of interviews that were conducted with STI practitioners.

These analyses culminated in a "position paper" prepared by the editors, that was provided to participating marketing theorists prior to the preparation of their conference papers. The purpose of this position paper, which serves as one of two introductory papers for this volume, was to familiarize marketing theorists who may not have worked in the STI field with the makeup, problems, and research opportunities in the field.

The conference was an open and "off-the-record" opportunity for STI marketing practitioners and marketing theorists to pose problems, propose solutions or studies addressing those problems, and critique each other's concepts and ideas. Discussion was extensive and sometimes heated. Yet, the two groups came, in a two-day period, to an appreciation of each other's skills and a greater awareness of the potential values of cooperation through research.

The participating marketing theorists were given the opportunity to revise their ideas based on the conference discussions. The papers in this volume represent those revisions.

This volume has the objective of extending the conference values to a broader audience. It is intended to demonstrate the potential utility of marketing theory to STI practitioners and the potential for research and application in the STI area to marketing researchers. The manner in which this is done is admittedly speculative. However, the speculation has been subjected to the scrutiny of a distinguished group of STI practitioners and marketing scholars and found to have merit.

As codirectors of the conference and coeditors of this volume, we would like to thank the conference participants for their time and for the energy they devoted to the undertaking. Special thanks are due to Dr. Joel Goldhar, formerly of the National Science Foundation, without whom neither the intellectual basis for the conference nor the conference itself would have come into being. Rohit Deshpande, a doctoral candidate at the Graduate School of Business of the University of Pittsburgh, also deserves recognition for his intellectual and administrative contributions, as does Dean H. J. Zoffer of the Graduate School of Business, who has provided the environment, leadership, and motivation that are necessary for the development and nurturing of innovative ideas and areas of research.

William R. King
Gerald Zaltman

Part 1

Introduction

Introduction

Information transfer is central to both science and technology. The importance of information transfer as an area of investigation is directly proportional to the volume of information to be disseminated and used. The volume of scientific and technical information (STI) is increasing very substantially. For example, the number of scientific and technical books published per year has risen from 3500 titles in 1960 to an estimated 16,500 titles in 1980. The number of other types of scientific and technical documents published yearly is also increasing. Moreover, the volume of information in some of these documents is also increasing. It is estimated that the average number of articles per scientific and technical journal will increase by 100 percent between 1970 and 1980.

Paralleling the substantial increases in volume is a rising increase in communication costs. In terms of constant dollars (1972) the investment per scientist or engineer for scientific and technical information increased from $2,500 in 1960 to $3,900 in 1975. This figure is expected to reach $4,200 by 1980 (King Research, Inc., 1977, pp. 3, 8, 11, 15).

It is likely that there will be continuing increases in both the annual quantity of published documents per scientist and engineer and the real cost per capita of generating and publishing the information contained in these documents. This continuing information and cost increase greatly influence the importance of information transfer, since information can acquire value only after it is transferred to users.

The significance of information transfer as a problem area

very much in need of attention was reflected several years ago in the selection of this topic as the major theme for the Annual Convention of the American Documentation Institute. (Of course, the importance of information transfer was recognized much earlier as well [De Solla Price, 1963].) Ten years later the feeling still prevails—perhaps even more strongly—that information transfer is a very serious problem (Fahey and King, 1977; Gerstenfeld and Berger, 1978; Kelly and Kranzberg, 1975; Gartner, 1975; Allen, 1977).

One constructive approach for addressing many issues in information transfer is through the development and use of better conceptualizations of the information transfer process. (See, for example, Murdock and Liston, 1968; Samuelson, 1974; Otten, 1970; Davis, 1978; and Glaser, 1977). Current conceptualizations of information transfer range from mathematical and statistical diffusion models such as those described by Goffman (1966; 1969), Zaltman and Kohler (1972), Nance, Korfhage and Bhat (1972) and simulation (Beres, Kohler and Zaltman, 1975) to highly specified logical flow models (Libbey and Zaltman, 1967; Murdock and Liston, 1967; Garvey et al., 1971; King Research, Inc., 1977; and Gerstenfeld, 1977).

However, Ganz and Goldhar (1975) suggest that extant information transfer conceptualizations are often too abstract and insufficiently prescriptive with regard to methods for improving information transfer to users. Samuelson (1974), in his review of various models, suggests that the present frontier lies in deriving a deeper understanding of what it is that current models or conceptualizations describe. This deeper understanding would permit more insights of a prescriptive nature into the problems of users (King and Cleland, 1975; King, 1978; Kochen, 1976).

Thus, any new conceptualizations to be used for understanding and improving information transfer should: (1) have prescriptive implications; (2) be based on theory derived from practice or applications; and (3) be derived from a discipline where the exchange or transfer process between producers/suppliers and users is the major focus of study. One area of academic theory and professional practice which meets these criteria is marketing. Marketing is "the process in which exchanges occur among persons and social groups" (Levy and Zaltman, 1975,

p. 26). It consists of the various factors involved in the creation, stimulation, facilitation, and valuation of transfers between two parties (Kotler, 1972). Thus, the tools and concepts of a transfer-oriented field should be relevant to information dissemination. Although this has been done in a limited way (King, 1978), the potentially greater applicability of marketing theory and practice to specific problems in the transfer of scientific and technical information, together with the importance of information transfer, offer exciting opportunities. However, these opportunities coexist with problems and challenges unlike those that have faced the marketers of consumer and industrial products.

This introductory section attempts to set the stage for the things to come by defining the state of the art in STI marketing and addressing some issues and opportunities that exist in this new field.

The introductory position paper by the editors attempts to summarize the existing situation with respect to the "STI system." This paper includes a description of the STI system and a "value added" analysis of the STI communication process along with other themes. The authors conclude this paper by identifying numerous problems and research opportunity areas. Many of these areas are discussed by other papers in this volume, and nearly all areas received substantial discussion during the conference. While the various areas identified are by no means exhaustive of important topics in STI marketing, there seemed to be a consensus that these are indeed critical issues very much in need of further research and remedial action.

The second paper in this section, by Joel Goldhar, describes some of the problems and speculates on the directions that research may take in resolving these problems and issues. An especially important concern raised by Goldhar is the need for information services to become much more user oriented. His belief that user needs and behaviors are often not considered in the design and implementation of information systems received considerable endorsement by conference participants. It is Goldhar's contention that the user orientation that marketing embodies, along with the tools and techniques of marketing, might fruitfully be applied by the STI industry.

W.R.K.
G.Z.

References

Allen, Thomas J. *Managing the Flow of Technology: Technology Transfer and the Dissemination of Technological Information Within the R&D Organization.* Cambridge, MA: MIT Press, 1977.

Beres, Mary Elizabeth; Barbara Marie Kohler; and Gerald Zaltman. "Communication Networks in a Developing Science: A Simulation of the Underlying Socio-Physical Structure." *Simulation and Games* 1 (March 1975):3-38.

Davis, Howard. "Knowledge Into Action: What Do We Know As Practitioners." Keynote address presented at the Research Utilization Conference, University of Pittsburgh, September 20-22, 1978.

De Solla Price, Derek. *Little Science, Big Science.* New York: Columbia University Press, 1963.

Fahey, Liam, and William R. King. "Environment Scanning in Corporate Planning." *Business Horizons*, August 1977.

Ganz, Carol, and Joel Goldhar. "The Role of Scientific Communication in the Process of Technological Information." *Information*, October 1975.

Gartner, Joseph. "Filling the Gap in the Information Revolution." *Proceedings of the American Society for Information Science*, 1975.

Garvey, William D.; Nan Lin; and C. E. Nelson. "Communication in the Physical and Social Sciences." *Science*, December 11, 1970.

Gerstenfeld, Arthur. Proposal submitted to the National Science Foundation, 1977.

Gerstenfeld, Arthur, and Paul D. Berger, "An Experimental Design for Improving the Transfer of Information to Scientists and Engineers." Paper presented at the American Institute of Decision Sciences Meeting, San Diego, March, 1978.

Glaser, Edward, et al. *Putting Knowledge to Use: A Distillation of the Literature Regarding Knowledge Transfer and Change.* Los Angeles: Human Interaction Research Institute and the Mental Health Services Development Branch, NIMH.

Goffman, William. "Mathematical Approach to the Spread of Scientific Ideas—The History of Mast Cell Research." *Nature* 212. 5061 (October 1966).

Goffman, William, and Kenneth S. Warren. "Dispersion of Papers Among Journals Based on a Mathematical Analysis of Two Diverse Medical Literatures." *Nature* 221. 5187 (March, 1969).

Kelly, P., and M. Kranzberg. *Technological Innovation: A Critical Review of Current Knowledge.* Vols. 1 and 2. Advanced Technology and Science Studies, Georgia Institute of Technology, 1975.

King Research, Inc. *A Chart Book of Indicators of Scientific and Technical*

Communication in the United States. Washington, D.C.: Superintendent of Documents, U.S. Government Printing Office, 1977.

King, William R. "Designing Useful Management Decision Support Systems." *Management Decision* 16. 4 (1978).

King, William R., and David I. Cleland. "The Design of Management Information Systems: An Information Analysis Approach." *Management Science*, November 1975.

Kochen, Manfred. "Can the Behavioral Sciences Contribute to the Foundations of Information Sciences?" *Proceedings of the American Society for Information Science*, 1976.

Kotler, Philip. "What Consumerism Means for Marketing." *Harvard Business Review*, May/June 1972.

Levy, Sidney J., and Gerald Zaltman. *Marketing Society and Conflict.* Englewood Cliffs, N.J.: Prentice-Hall, 1975.

Libbey, M., and G. Zaltman. "The Role and Distribution of Written Informal Communication in Theoretical High Energy Physics." New York: American Institute of Physics, Report No. AID/SDD-1 (rev.) Report No. NYO-3732-1 (Rev.), 1967.

Murdock, John W., and David M. Liston, Jr. "A General Model of Information Transfer: Theme Paper 1968 Annual Convention." *American Documentation*, October 1967, pp. 197-208.

Nance, Richard E.; Robert R. Korfhage, and U. Naryan Bhat. "Information Networks: Definitions and Message Transfer Models." *Journal of the American Society for Information Science* 23. 4 (July/August 1972):237-247.

Otten, Klaus W. "Information and Communication: A Conceptual Model as a Framework for Development of Theories of Information." In Anthony Debons and William J. Cameron, eds., *Perspectives in Information Science*, pp. 127-148. Woohoff-Leyden, 1970.

Samuelson, Kjell. "Information Models and Theories—A Synthesizing Approach." in Anthony Debons, ed., *Information Science: Search for Identity*. pp. 47-67. New York: Marcel Dekker, Inc., 1974.

Zaltman, Gerlad, and Barbara Marie Kohler. "The Dissemination of Task and Socioemotional Information in an International Community of Scientists." *Journal of the American Society for Information Science* 23. 4 (July/August 1972): pp. 225-236.

1
Marketing Scientific and Technical Information

William R. King
Gerald Zaltman

The process of creating and disseminating scientific and technical information (STI) can be likened to that of producing and distributing a product or service. Although this view of STI as a commodity to be marketed is a natural conceptualization for marketing scholars and practitioners, it is not one that has been extensively applied to STI policy or research.

Objectives of the Paper

This "position paper" has the objective of describing the state of the art in terms of the existing STI system, current practices in the marketing of STI, problems and research opportunities in the area, as well as the nature of an STI marketing conference that was held at the University of Pittsburgh. The paper draws heavily on the previous work of others as well as on interviews and assessments which were made by the authors during the preparatory phase of the conference.

The STI Marketing Conference

The primary objective of the conference was the *evaluation of the applicability of a variety of areas of marketing theory to the marketing of scientific and technical information*. This evaluation was conducted by a panel of experts in various areas of marketing theory who prepared position papers relating their area of marketing expertise to the problems of marketing scientific and technical information, and through their involvement in the conference, participated in the critical analysis and refinement of their ideas.

Overview of STI

The communication of STI is readily and intuitively recognized to be one of the most important determinants of scientific and technical progress. The history of science and technology is replete with instances of progress that are based on information describing the accomplishments of other scientists and technologists. Unfortunately, until very recently, little has been done to assess the STI communications process and to work toward the achievement of communications improvements.

The reasons for this lack of research and study concerning STI are manifold. First, information of any variety has only recently come to be thought of as being susceptible to study using the techniques that have been applied elsewhere. Perhaps more important, however, is the simple recognition that "more communication is not necessarily better than less communication." Ackoff[1] has made this point in the context of management information systems and others have found empirical evidence of this in the STI context, at least within some narrowly defined sets such as external communications vs. internal communications or research vs. development efforts.

Perceptions such as this one have led to increasing thought and research concerning the processes through which STI users identify, acquire, and use relevant information. Since these processes are so much like those undergone by a consumer in selecting and purchasing a product or service, the potential applicability of marketing ideas and theory is directly suggested.

The Magnitude and Significance of STI

The total resources (funds and manpower) expended for STI communication in the U.S. are estimated at $9.4 billion for 1975.[2] "This includes the costs incurred by authors, publishers, libraries, and secondary services in the production and use of scientific and technical books, journals, reports, and other publications."[3]

These resource expenditures are growing at a faster rate than any of the indicators that might be expected to "track" them—

e.g., GNP, R&D funding levels, number of scientists and engineers, etc. Despite a recent leveling off in the demand for scientists and a corresponding leveling-off in salaries, these increases are expected to continue at a real annual rate of about 3 percent through 1980.[4]

A major portion of these overall STI resource expenditures is for scientific and technical journals (about 63 percent), with books accounting for about 28 percent of the total (in 1960). About 60-70 percent of the total is accounted for by individual scientists and technologists in writing, editing, reviewing, and reading literature. Libraries represent about 25 percent of the total and publishers represent about 10 percent.

The STI System

The current system for STI communications is really a collection of uncoordinated and independent entities that loosely link information producers and users. There is, in fact, no overall system in the usual sense in which the term is used.

Indeed, even the conceptualization of a single STI system that encompasses the physical, social, and engineering sciences may be too great an abstraction to be useful. However, since such an abstract model is commonly used, we shall discuss "the system" in such terms here. In doing so, we shall give attention to those aspects of the system that would appear to be most significantly affected by differences in the various scientific contexts.

The difficulties in studying such a "nonsystem" are even greater because STI producers and users are often the same people and organizations—i.e., individuals and organizations are simultaneously users and producers of information.

STI System Actors and Their Motivations

The primary actors in the STI system are scientists and engineers who produce information and who, in turn, require information to support their own research and development activities. The motivations of these actors are multidimensional and usually involve (on the producing side): (*a*) a perceived obligation to the scientific and professional community, (*b*) an organizational obligation (such as the "publish or perish" syndrome), and (*c*) a desire to achieve personal visibility and to have one's accomplishments recorded and widely recognized.

In their STI user role, the actors are primarily motivated by a desire to achieve scientific or technical progress on a cost-effective basis. This progress may be measured in abstract terms of "achieving greater understanding" or in more practical terms such as "new products created." In either case, the desire is to achieve progress by building on the work and results already achieved by others.

Other actors in the system are the libraries, publishers, and secondary communication services and "organizations" who act on behalf of individual users and producers of information—e.g., companies, research laboratories, etc.

STI System Messages

Although most discussions of STI tend to focus on journal articles, there are a variety of formats that STI messages take on. Ackoff et al.[5] distinguish three levels of scientific messages of both the formal and informal varieties as shown in Table 1.1. These various messages entail both primary scientific and technical information and "information about information" (secondary or tertiary messages that reference, summarize, or analyze primary information).

While the output of scientific research is normally recorded in written form, the output of technological development may well be a product or a process that is documented only partially through operating handbooks and maintenance manuals. To some degree this is because of the objectives and nature of the technology process, but it also reflects secrecy and confidentiality considerations that may be imposed by the government or a private company.

The STI Communications Process

Ackoff et al. have described the STI communications process (for recorded "messages") as consisting of four phases[6]:

1. Production   3. Acquisition
2. Dissemination   4. Use

TABLE 1.1

CLASSIFICATION OF SCIENTIFIC MESSAGES

	FORMAL	INFORMAL
PRIMARY	Journal articles, monographs, dissertations, proceedings of meetings, reports, preprints, patents	Meetings, seminars, letters, conversations
SECONDARY	Bibliographies, listings, indexes, abstract journals, SDI services, dictionaries, catalogs, newsletters	Letters, conversations
TERTIARY	Critical reviews, state-of-the-art surveys, review books, yearbooks, handbooks, textbooks, encyclopedias	Lectures (classes), conferences, symposia, letters, conversations

One difficulty in studying the STI system is that several of these conceptual phases may be operationally combined so as to be indistinguishable. However, the taxonomy serves as a useful basis for describing the STI communications process.

Production

The production of scientific information is itself a process that proceeds through a number of successively more formal stages. The process may begin with the scientist preparing notes, discussing them with colleagues, and then preparing and circulating a draft within his own organization or narrow circle of professional colleagues. Then a revised version may be presented at a formal scientific meeting before another revision, based on the cumulative feedback, is submitted for publication. The style, volume, and other constraints of the publication, coupled with referees' comments, may require further revisions to be made before publication.

A good deal of study has been performed on the various stages of the production cycle and the time that elapses between the initial step and publication. For instance, Garvey, Lin, and Nelson[7] have compared these stages and times for the physical and social sciences and found that the process takes 80 percent longer in the social sciences and that the stages are not so rigidly organized as in the physical sciences.

The most significant delays in the production process occur between formal submission and publication. This varies from field to field and journal to journal, but delays of 15-24 months are not uncommon. This delay is caused by refereeing and the mechanics of the printing process. New technological developments would appear to offer significant promise of reducing this mechanical delay element.

Dissemination

The dissemination phase of the STI communications process involves those methods by which primary, secondary, and tertiary messages are made available to potential users. Thus, this phase deals with the dissemination of both information and "information about information" (e.g., references, bibliographic listings, etc.).

The physical distribution of recorded messages may be performed directly, as in the sending of a paper to a colleague, or indirectly, through publishers, libraries, or bookstores. Scientists tend to receive more unsolicited documents than solicited ones, making filtering of great importance to the potential user.

Technologists, as opposed to scientists, tend to make greater use of others working within their own organization. The concept of a "gatekeeper"—an individual who is recognized as a source of information about information in a particular field, specialty, or related area—has been described[8] and prescriptively implemented by firms such as U.S. Steel as a device for improved dissemination. In any case, physical and organizational proximity comes into play as a major determinant of technological communication.

Even among scientists, informal "colleges" (networks) serve as effective and psychologically satisfying ways of communications. Such colleges tend to be closed clubs that facilitate communications among the members while discouraging the addition of new members.

Publication delays, as well as the high rejection rates for some journals, has led to increased use of reference messages. This information about information may take the form of journal listings of working papers ("preprints") that are available from the author, journal listings of papers accepted for future issues, organizational mailings of working paper listings, newsletters, and so on. Among the new "reference products" that have achieved great commercial success in this regard is *Current Contents*, a publication of the Institute for Scientific Information (ISI) that consists of weekly compilations of recent journal tables of contents.

There has been enormous recent growth in commercial "information brokerage"—i.e., firms specializing in information about information sources. Many of these services involve computerized data bases that facilitate search processes paid for by potential information users. Of course, duplication and lack of standardization hamper the cost-effectiveness of these services. This inevitably results in redundant and overlapping information being provided to potential users.

Acquisition

The STI acquisition process has been extensively studied to determine the relative importance of various information sources. Ackoff et al. describe the results of these studied in terms of four "significant findings":[9]

1. A relatively small core of journals and other publications is used to satisfy most requests for information. This raises questions about the growing number of journals that are available.
2. Formal acquisition systems are not used as heavily as informal ones if they do not have the same desirable characteristics as the informal: ease of use, accessibility, and responsiveness.
3. Channel use varies with the type of search being conducted and the place of research. The basic researcher looking for theoretical information will be likely to use formal channels. Researchers in universities consider journals to be

very important, while those in private firms rely more heavily on informal channels.
4. Researchers have a variety of needs that change throughout their work. Formal channels are used most heavily by researchers during the idea-generating state of research, while informal channels are useful in amplifying and clarifying information and defining problems.

These studies have not been conducted with the objective of STI system improvement. Therefore, the improvements they suggest have been neither systematically developed nor implemented.

Use

STI use involves the extraction of content from a message to meet a need. Various kinds of needs have been identified by Orr as a result of his study of biologists[10] :

1. Regular needs
    Current awareness
    Everyday reference (specific information for the support of current work)
    Personal stimulation
    Personal feedback (to obtain reaction to one's own work)
2. Episodic needs
    Retrospective search—Exhaustive ("all" prior work)
    Retrospective search—Limited
    Instruction
    Consultation

User needs studies have tended to focus on *document* needs, as opposed to *information* (substance) needs. Some of the varieties of needs noted above are best served by documents (e.g., awareness, stimulation, and factual searches) while informal channels best serve other needs (e.g., feedback).

Scientists and technologists have not been found to spend a great deal of time in reading, although about 20% of their time may be spent in scientific communication.[11] However, more time is spent in actual reading if literature is made readily available (e.g., at the desk).

Martin and Ackoff have also found that abstracts are frequently used as a substitute for, rather than as a guide to, primary information.[12] This fact, coupled with recent court decisions that will restrict photocopying, calls into question the entire relationship between the primary STI system and the subsystem, which encompasses "information about information".

"Value Added" Analysis
of the STI Communications Process

Another perspective on the STI communications process involves assessing the "value added" to this process by the various primary, secondary, and tertiary components identified in Table 1.1. This perspective augments the four-phase approach just discussed. Each component in the STI communications process may be evaluated in terms of its present and potential role in providing various kinds of value. We shall identify several dimensions of "value" below.

- *Identification of user needs.* Does—or could—a component provide insight into the kinds of information desired by users or potential users?
- *Creation/production of data.* Does a component stimulate the development of new data and ideas? For example, symposia might stimulate the creation of new STI.
- *Creation of awareness of data.* How many users become aware of STI only as a result of a component? How much earlier is awareness established among users as a result of a component such as preprints?
- *Quality control: validity of data.* Does a component such as refereed journals impact the quality of the data transmitted?
- *Quality control: potential usefulness of data.* How helpful is a component in enhancing the usefulness of data? Does a component such as an abstract journal allow the practical relevance of data to become evident?
- *Storage.* How valuable is a component for warehousing purposes?

- *Retrieval.* How easily can data be recalled via a particular component?
- *Translation of data into implications.* Does the nature of a component permit a statement of the way in which the data are important? Can bibliographies be retrieved on the basis of implications?
- *Translation into action.* Does a component permit statements about the actual use of the data once its relevance is determined?
- *Feedback.* Does the component permit feedback from the user to the disseminator and creator of the data? How many journal channels must be altered to facilitate feedback?

The above are only a sample of the various dimensions of "value" that a component of the STI communications process may offer. The various dimensions are probably differentially salient or important in each of the four stages of the STI communications process. For example, the identification of user needs is especially important for the production phase. The creation of awareness is clearly important at the dissemination stage as are the values of quality controls. Storage and retrieval are important at the acquisition phase. Translation values are particularly relevant at the user phase.

STI System Usage, Economics, and Outlook

The economics of the STI system constitute an important base point for any analysis or prescription relative to the system.[13]

Economics and Usage of the Individual Scientist

On a "per scientist or engineer" basis, total STI resource expenditures increased from about $2000 to $3000 in real terms between 1960 and 1974. This (real) level is anticipated to hold about constant through 1980.

However, the largest element of expenditure has, and will continue to be, the *user*, who expends both time and funds in reviewing, identifying, and using STI. This represented about

50% of total expenditures in 190, but is expected to decline to about 40% of the total in 1980. This anticipated decrease is attributable to an expected dampening in the number of employed scientists and salary levels in the sciences.

Producers of STI (authors) accounted for 17% of total expenditures in 1960; this proportion is expected to increase to 23% by 1980. Correspondingly, the productivity of scientists has generally declined since 1960 (from one article per 10.9 scientists in 1960 to one per 13.1 scientists in 1974), but there are substantial variations among various scientific fields.

Economics and Usage of STI Literature

Scientists and engineers rely more heavily on journal issues and references cited in them than on formal indexing or abstracting services for initially identifying journal articles that they cite in their own articles. As a consequence, journals constitute an important element of the STI system.

The economic outlook for journals is more positive than for some other elements of the system. The number of scientific and technical journals has not increased dramatically since 1960, with the steady 2% increase reflecting almost perfectly the steady growth in the number of scientists.

The number of journal subscriptions has increased on the average over this period, although individual journals have experienced dramatic subscription decreases. This has occurred while journal prices (real) have increased slightly. All of these trends are expected to continue, with some possible moderation, through 1980.

Book publishing, on the other hand, has a less positive outlook. Book titles published have increased nearly fivefold in the 1960-1974 period. Author productivity has increased from one book per 343 scientists to one per 137 scientists. Concurrently, the number of book copies per title has dropped more than 50% during the period and, even though prices have risen dramatically, the revenue per title has declined sharply. Book copy sales per scientist have been relatively stable at 6-8 copies while the number of titles has increased and the average number of copies sold per title has decreased.

The forecast is for a continued belt tightening in the book

publishing industry, with increased prices, stringent manuscript prescreening, and severe cost reductions.

Report literature, conference proceedings, and doctoral dissertations have also increased substantially in the 1960-1974 period. The *use* of these forms of communication has also increased, at least as indicated by sales of the National Technical Information Service (NTIS) and the Government Printing Office (GPO). Microfilm copies of reports have increased more rapidly than paper copies, and because of continued relative price increases for paper, the trend is expected to continue.

Economics and Usage of Libraries and Other STI Agents

Because the production and use phases of STI communications usually occurs after some time intervals, libraries and other institutions serve as both communicators and repositories of information.

Between 1960 and 1974, STI library expenditures expanded 200-250% in real terms. However, growth peaked in 1972 and is not expected to increase in the period to 1980. This will require libraries to increase their technological content (probably through increased computerization) and their effectiveness-cost ratios (probably through such devices as computer networks).

Libraries, as well as other abstracting and indexing services, probably account for over 20% of the article *identifications* made by scientists. These services are projected to increase substantially by 1980 as greater coverage is accomplished and the potential for computerized search of bibliographic data bases is expanded.

STI "Problems" and Research Opportunity Areas

Since a primary purpose of the STI Marketing Conference was the identification of "high impact" areas for the application of marketing theory, it is appropriate to close this paper by preliminarily suggesting some problem and research opportunity areas that have been identified[14]:

1. Development of cost models and analytic capabilities—to

permit the improved prediction of publication and marketing costs.

2. Organizational mechanisms for cooperative publishing—to permit cross fertilization as well as the achievement of cost economies.

3. "Product" redesign—to achieve greater utility from scientific messages (such as "use statements" included in articles).

4. Improved systems for selective information dissemination—to utilize "user profiles" so that relevant information can be automatically and selectively provided.

5. Achievement of greater timeliness of research information—such as through the publication of a brief note indicating current directions of author along with an article reporting "results".

6. Optimization of physical availability of STI—particularly within a given organization.

7. Improved study of substantive information needs as opposed to documentary needs.

8. STI market segmentation and differential needs assessment.

9. Impact of organizational "style" and policy on STI use and production.

10. Invasion of privacy issues (e.g., analyzing who asks for what information).

11. Special techniques for providing STI services to small firms.

12. Impact of confidentiality and proprietary policies on STI system.

13. Techniques for reducing publication delays.

14. "Filtering" techniques to permit individual users to better fulfill needs.

15. Positive and negative values of "gatekeepers."

16. Improvements in invisible college networks.

17. Organization of information brokerage market.

18. Study of user needs in information terms rather than in document terms.

19. Structure and linkages of STI subsystems dealing with primary information and "information about information."

Notes

1. Russell L. Ackoff, "Management Misinformation Systems," *Management Science*, Vol. 14, No. 4, December, 1967, pp. B147-B156.

2. National Science Foundation, "Statistical Indicators of Scientific and Technical Communication: 1960-1980," Report prepared by King Research, Inc., 1976, p. 13.

3. Ibid.

4. Ibid., p. 14.

5. Russell L. Ackoff, et al., "Scientific Communication and Technology Transfer system," Vol. 2, Report of the University of Pennsylvania for the National Science Foundation, NSF-GN-41883, 1975.

6. Ibid.

7. William D. Garvey, Nair Lin, and Carnot E. Nelson, "Some Comparisons of Communications Activities in the Physical and Social Sciences," in C. D. Nelson and D. K. Pollack, eds., *Communications Among Scientists and Engineers*, Health Lexington Books, 1970.

8. For instance, see T. J. Allen, J. M. Piepmeir, and S. Cooney, "The International Technology Gatekeeper," *CP&E*, 1972.

9. Ackoff, et al., pp. 7-49.

10. Richard H. Orr, "Milieu of U.K. Secondary Services 1978-1983," Report for the Committee on Biological Information, June 4, 1973 (Mimeo).

11. Miles W. Martin and Russell L. Ackoff, "The Dissemination and Use of Recorded Scientific Information," *Management Science*, Vol. 9, 1963, pp. 322-336.

12. Ibid.

13. All statistics and estimates in this section are adapted from National Science Foundation, Chaps. 2, 3, 4, and 5.

14. From the published literature of STI as well as on the basis of informal interviews of users and disseminators conducted by the authors.

2

STI Dissemination: Issues and Opportunities

Joel Goldhar

At the time of this workshop, the charter of the Division of Science Information (DSI) of the National Science Foundation (NSF) addressed the problem of improving the quality, effectiveness, and efficiency of the communications and information infrastructure in research and engineering. Although there have been some changes in the direction of DSI, the issues raised in this volume are critical ones for the growth and enhancement of science and technology as it depends upon the efficient and effective transfer of information.

It has long been recognized that the processing and transfer of information is an essential element of research. To a somewhat lesser extent, it is also essential to engineering. Information is the output of science, and if one defines the formulation of new technology to be a form of information, it is also the "product" of engineering. Moreover, the output of one individual's or organization's research is the input to the scientific activities of others. Thus, an essential issue of STI is: How can we effectively and efficiently distribute information that is the output of one technical effort to provide the inputs for another technical effort when these efforts are separated in time, space, and "culture"?[1] This is a particularly difficult issue when multiple scientific disciplines are involved, or when the transfer of information is between basic science and engineering applications.

The traditional STI infrastructure has developed to deal primarily with transfers of information in the time dimension. A physical chemist today can rather easily determine what previous physical chemists have accomplished. He can, with somewhat more difficulty, find what other varieties of chemists have previously accomplished, particularly if he uses informal systems as well as formal ones.

In the past twenty years, attention has also been devoted to transferring information across the space barrier. Scientists today have a fairly good system for learning what is going on in other nations in their fields, in part because of the continued and historical internationalization of science.

However, we have not yet very effectively addressed the "culture gap"—the transfer of information among disciplines and between science and engineering. Today, scientists and engineers find themselves working on problems that are inherently interdisciplinary. They can no longer rely on their personal education, experience, and local information bases. They must, in order to participate effectively in modern problem-focused research, have access to information that does not appear in the literature of their own specialty. Often, engineers cannot even rely on engineering literature in other fields, but must access basic scientific information in order to solve applied problems.

For this to be feasible, effective, and efficient, some changes must occur in the STI infrastructure—the primary journals, the abstracting and indexing services, the libraries, corporate information centers, public libraries, etc. There is a need for a new generation of information systems and services for the dissemination of STI. This has already begun, but I believe that it needs to increase in intensity in the coming decade.

This process began with judgments made by various distinguished individuals and committees who analyzed the state of science and engineering and concluded that, despite the magnitude and dynamism of the existing STI system, users did not feel they were well served.[2] A little closer analysis showed that many information services are not user oriented, that users are not involved in the design of the information system, and that many of the services don't know how their information is really used by scientists and engineers. The traditional approach had been to group uses by discipline and not to recognize the differential needs of various users within each discipline. A newer and more useful concept is that user needs may be very different if the user is advancing knowledge, or developing new products or processes, or establishing policy.

NSF's Division of Science Information Service has focused upon making the STI industry more effective and more user oriented. My own assessment of the STI infrastructure is one of a noninnovative, fragmented, capital-deficient industry. This characterization reflects a need for risk capital to permit the development of new products and services, experimentation, and market research. It also reflects a need to overcome a history of nonaggressiveness and aloofness from users. These problems are additionally complicated by the mix of for-profit, not-for-profit, and government run information services.

Changing this situation requires visibility for the information "business." Unfortunately, STI is like other intangibles: it is not valued highly until it is unobtainable. Thus, the "great men" of science, who mostly rely on informal communications rather than on formal ones, often do not give high priority to STI. Those who do—the engineers, graduate students, and laboratory scientists—are neither vocal nor visible. So there is a basic image problem which must be addressed.

We can begin to create change in information transfer by concentrating on three leverage points: First, we can deal with the information services by helping managers to know their users and markets better and encourage them to experiment with both technological and organizational innovation; second, we can deal with the user to make him more aware of what is available and more willing to cope with the added uncertainty and problems that additional information may bring into his life; thirdly, we can focus on the managers of organizations within which information users operate.

We can address information service managers by providing motivation amd funding for initial trials of innovations—including innovations that are organizational and institutional as well as the application of modern information technology. The user can be addressed through programs of familiarization with the information and services that are available. However, we must also deal with the basic fact that more information creates psychological dissonance. Users may react defensively to new information. They know that more information implies additional work, uncertainty, and the necessity to seek even more information. There is a loyalty to tried and true information sources and methodologies. Thus, to address the user requires creating changes in behavior and thinking patterns, a very difficult thing to do. If it is not done, however, new systems for disseminating STI will not be effective. If information is delivered *to* individuals, but is not processed *by* them to impact on their work, it has no real value.

The third leverage point—R&D managers—has a great deal to do with the willingness of potential users to accept information from external sources, to use it, to share it, and to communicate with others. We have begun to learn that it is important to integrate formal and informal information systems. The trend for a while was to think of formal and informal communications as substitutable; if you didn't use one, the other could be used to achieve similar ends. Now, we recognize the formal and informal as part of an overall system. And it is the R&D managers who must be made aware of this, so that they can establish systems, procedures, and organizations to ensure that information is captured, used, and communicated as widely as is desirable.

This conference and the volume of papers it has produced are a part of the overall NSF program to improve the STI system. A large part of that improvement will come, in my judgment, through the creation of a marketing orientation on the part of the STI services. There is a great need to identify users, to segment the market, to learn about user behavior, and to design services that "fit."

Marketing has done these things in other industries and with other products and services. However, they have not yet been applied to the STI case. In those instances where marketing has come into play, it has often been relatively unsophisticated. For instance, much of the industry views marketing as equivalent to a sales orientation, and market research is often done as an extension of what the trade publications do to find out who their readers are. The STI industry is just beginning to address marketing issues in a more sophisticated way. To do this well is not an easy task, in part because of the history of the industry, its tradition, and its structure, as well as the inherent nature of the information commodity.

Some think of information as a pure service business: you have it, you sell it, and you still have it to sell. Others use the consulting analogy: you have it, you sell it, and you then have more of it because you've learned something from the client contact. But, it is a more complex situation than either of these analogies imply. Information is a unique product with a peculiar combination of difficult-to-manage characteristics. It is not a consumer durable, although it has some of those attributes. It is not a packaged good, although some of its elements, such as books and journals, are. It is not a pure service business, but some elements of it are. It is some combination of each of these, with an industry structure that makes it difficult to separate the parts.

Yet with all of these difficulties, information is a commodity in which marketing can create a much larger proportion of the value of the product than with most other products. Information has much of its value precisely because of distribution and potential users' awareness of it. The timeliness, distribution, and format of information can change its value from nothing to infinite. Marketing is the process of creating value through the creation of time, place, and form utilities. Thus, it is natural to expect marketing theory to be applicable to the information commodity. These conference papers should go a long way toward defining just how that might best be accomplished.

Notes

1. Culture is defined as everything else but in this context refers especially to differences among disciplines, between basic and applied activities, and between academic and industrial working environments.

2. For further reference see *Scientific and Technical Communication*, Washington, D.C.: National Academy of Sciences, 1969. Later studies essentially confirmed and updated the conclusions of the *SATCOM* report.

Part 2

The STI User and Market

Introduction

An important concept of modern marketing thought has to do with the adoption, by marketers, of a "consumer orientation." This simple idea says that products and services should be developed and marketed to fulfill user needs. This is in contrast to a "product orientation" in which a producer develops a product or service often on the basis of what is technically feasible and only then determines whether a user need exists that can be satisfied by the particular product or service that has been developed.

Despite the simplicity of the idea, it has often been violated in practice through the development of products in isolation from realistic understandings of consumers, their purchase-decision processes, and their motivations for making purchases.

The three papers in this section deal with users of and the market for scientific and technical information. Collectively, they present several useful ideas for including the user in all phases of the information production and dissemination processes. The user may be an individual within an R&D department, the entire firm, or collections of firms grouped in unique ways. These papers considered together express a point made elsewhere: "Marketing takes the focus off the product and puts its on the user's needs. Products are developed from the user's point of view, not the producer's" (Kotler et al., 1977). This does not mean that the producer and/or disseminator of scientific and technical information should be passive or uncreative and only originate ideas for information or information transfer systems after direct or indirect consultation with users. Rather, once an idea is originated, say by an indexing service or journal, it should be further developed with users in mind (Terrant and Garson, 1977).

Robert Rothberg deals with the firm as a user of scientific and technical information. He attempts to define and to structure the motivations and processes by which firms use information to help them in facing an uncertain future. Unlike many who view STI to be useful only in a narrow and highly technical sense and only to a group of individuals within the firm who have specialized technical interests, he views STI to be essential for a wide variety of the firm's strategic purposes.

Jagdish Sheth's paper develops a model of STI consumer behavior, in terms that are related to models of consumer behavior developed in other contexts. The objective of his model is to enhance understanding of STI user motivations and processes. Such understandings should facilitate the incorporation of user considerations into the makeup of the STI "product" and into the systems and processes used to disseminate STI.

The third paper in this section, by Yoram Wind and Robert Thomas, applies market segmentation concepts to the STI market. Their emphasis is on ways of defining and addressing "the" market in a fashion that will enhance the efficiency and effectiveness of the STI dissemination system. In addition to addressing the applicability of these ideas to STI, Wind and Thomas focus on some unresolved issues in the market segmentation area that may affect the application of these ideas in the STI context.

References

Kotler, Philip, Bobby J. Calder, Brian Sternthal, and Alice Tybout. "A Marketing Approach to the Development and Dissemination of Educational Products." *In* Michael Radnor, Durward Hofler, and Robert Rich, eds., *Information Dissemination and Exchange for Educational Innovations*, Part 2, p. 7. Evanston, IL: Northwestern University,

Center for the Interdisciplinary Study of Science and Technology, 1977.

Terrant, Seldon W., and Lorrin R. Garson. *Evaluation of a Dual Journal Concept.* Washington, D.C.: American Chemical Society, 1977.

W.R.K.
G.Z.

3

STI Acquisition and the Firm

Robert R. Rothberg

In recent years there has been a virtual explosion in scientific and technological knowledge. Informed observers estimate, for example, that the number of technical journals published throughout the world today exceeds 100,000 and that the total body of technical information is now doubling every ten years.

It is impossible for a researcher using conventional methods to inspect more than a small fraction of the scientific and technological information (STI) bearing on a particular topic. Given this gap between knowledge and awareness, as well as the high and rising costs of research, it has become imperative that better methods be developed for the collection, sorting, and dissemination of STI. This concern is manifested most clearly at the level of the firm, where acceptance of better techniques requires that they not only be cost-effective but compatible with current information acquisition procedures as well.

The Firm's Perspective

Scientific and technological information is consciously acquired by a firm for a purpose: to help it survive and, if possible, to grow. In order to survive and grow, the firm must know what the success requirements are in its present and potential product markets, what differentiates it from present and potential competitors, what constraints it must deal with in the form of government regulations and internal finances, and what resources, knowledge, and skills it must obtain to counter incipient threats or take advantage of potential opportunities.

The position taken here is that STI is acquired by the firm in order to help it identify or clarify the nature of these threats or opportunities as well as to facilitate invention and innovation. Thus, the acquisition of STI cannot be meaningfully examined in isolation. It must be based on an appreciation of the broader objectives and constraints of the enterprise and integrated with the performance and information acquisition activities of other business functions. Collectively, these set the boundaries and priorities for STI search.

Strategic planning provides a useful framework for analyzing this acquisition process. This refers to the specification of corporate objectives and goals and the development of time-phased plans of action for identifying and evaluating threats and opportunities, a *structure* for mobilizing and allocating resources, and a *system* for monitoring and controlling plans as they are put into effect.

Direction

The environment in which the firm must compete is not static but changing, and apparently changing faster with the passage of time. The firm must monitor both its microenvironment and its macroenvironment if it wishes to anticipate threats and identify opportunities in sufficient time to take effective action.

Two especially important kinds of information can be obtained from each environmental sector, the first pertaining to existing or potential demand and the second pertaining to science and technology. Because the distinction between these two types of information is often quite hazy, it pays to have marketers scanning the environment who can appreciate the technical significance of what they discover, and vice versa.

The Microenvironment

First, consider the information obtainable from the microenvironment. This consists of the firm and its competitors, their suppliers, and their distributors and customer/users.

Two kinds of information search are carried out here, intelligence gathering and formal investigation. The former includes conditioned viewing and other relatively unstructured efforts to obtain information. The latter refers to a deliberate effort, usually following a preestablished plan, to obtain somewhat more specific information of interest or use to the firm. Although the same kinds of information are frequently obtainable from either mode of search, there are tradeoffs to be considered between the cost and quality of the information obtained.

The marketing information one can generate in the microenvironment is outside the scope of this paper. On the scientific and technological side of the ledger, however, much of what can be obtained can be referred to as "events" or "happenings" as opposed to "prescriptions" or "recipes." The former often signal the need for some kind of action, including the need for further investigation. The latter are more often the product of such investigations. They help to clarify the kind of action that can and should be taken.

Table 3.1 illustrates some of the kinds of "event" and "prescription" information that can be obtained from the microenvironment.

This table illustrates several important points about the nature of STI search. First, news of "events" can strongly affect the priorities assigned to more formal investigation. The news in question may be picked up directly by those charged with the primary responsibility for STI acquisition, such as R&D personnel, or relayed to such individuals from other parts of the organization, such as sales or applications engineering. The source (and power) of the internal signal frequently has an important bearing on the use that is made of it. Second, event and prescription information can be obtained from outside sources in a variety of ways, including the perusal of published material, personal conversation or correspondence, or as a consequence of formal interfirm agreements involving such things as licensing, cosponsored research, or mergers and acquisitions. (Interfirm agreements are an important STI transfer mechanism that will not be considered in detail in this paper.)

TABLE 3.1

EXAMPLES OF SCIENTIFIC AND TECHNOLOGICAL

INFORMATION OBTAINABLE FROM THE MICROENVIRONMENT

Source	Events	Prescriptions
A. Suppliers	1. New developments in materials or technology	1. Technical assistance in the proper use of their own products
	2. News concerning competitor requests for technical assistance	2. Output from user-commissioned research
		3. Joint research and development projects
B. Competitors	1. Changes in materials, technology and product design (from formal analysis or informal inspection of competitive offerings).	1. Patent applications by competitors
	2. Plans, accomplishments, and problems of a scientific or technological character (from public announcements or published reports)	2. Technical papers read at professional meetings or reported in professional journals by competitor personnel
		3. Suggestions from scientists and engineers newly hired away from competitors
C. Marketing Intermediaries and Final Users	1. Complaints, suggestions, and observations concerning product performance, materials and/or assembly techniques	1. Requests to produce specific pieces of equipment according to customer-supplied specifications, instructions and/or blueprints
	2. Product cost/effectiveness evaluations under actual use conditions	

The Macroenvironment

The macroenvironment of the firm includes such general forces as demography, economics, culture, government, and technology. Although these forces are essentially beyond the firm's influence or control, changes in their direction or intensity can have a considerable impact on the realization of the firm's long-term objectives. If the firm is to anticipate and adjust to these forces in a strategic sense, it has to acquire information concerning their implications for its present and potential markets and technology.

Insofar as marketing information is concerned, formal investigation plays a far more important role than intelligence gathering because the considerations at issue fall outside the scope of day-to-day marketing routine. These investigations tend to concentrate on the implications of macroenvironmental trends for the size and composition of the firm's present and potential product markets. The results of such studies frequently have a very strong bearing on the future directions of research and development activity.

On the scientific and technological side, the search for information from the macroenvironment is a continuous activity, embracing both intelligence gathering and formal investigation. It is very difficult to separate microenvironment activity from macroenvironment activity in this respect. Unlike the microenvironment, however, where emphasis is given to specific events in the macroenvironment, the firm tends to focus on the search for prescription types of STI. At this juncture, problems have already been defined in general terms and assistance is sought for their solution.

Thus far the direction taken by STI search has been discussed in terms of the environments to be studied, the modes of investigation, and the forms in which meaningful information is likely to be obtained. Little reference has been made, however, to the information acquisition problems created by rapidly expanding scientific and technological knowledge.

The Information Gap

Most firms with active and aggressive R&D departments are encountering serious difficulties trying to keep up with the flood of new STI as this is manifested in a wide variety of published source material.

From the standpoint of the firm's scientists and engineers, the main concerns are cost, availability, and mode of presentation. Once the most likely and/or readily available materials have been consulted, these individuals have to make some hard choices between the benefits and costs of additional search. It is difficult at the best of times to estimate the likelihood and magnitude of these benefits. Costs, unfortunately, both in time and money, are all too real. The cost side of the equation is further complicated by the researcher's need in many instances to obtain authorization from higher authority for all but the most modest outside purchases of STI.

One is led to conclude that the search for STI is conditioned far more heavily by the time, permission requirement, and budgetary constraints on the prospective user than any rational assessment of costs and benefits would first seem to indicate.

To be sure, a great many shortcuts have been developed to facilitate STI transfer. These range from various discipline and problem-oriented indices and abstracts to document repositories and highly-sophisticated, computer-based information retrieval systems. However, these shortcuts in many ways are accompanied by their own sets of problems.

Retrieval systems. Computer-based retrieval systems, in particular, would seem to hold out the promise of greatly expanded search capability at relatively low cost. Many such systems have been set up to operate as independent or quasi-independent STI utilities, servicing the needs of many firms simultaneously. Given the magnitude of the need it would be reasonable to assume that such systems are having a major impact on the STI acquisition practices of the research-oriented firm. This may be true. It is equally true, however, that the demand for the services of such systems has fallen considerably short of systems-sponsor expectations.

Part of the problem might be attributable to the inherent limitations of any electronic data processing-based retrieval system. Given the array of materials available as original input, updating requirements, and the variety of information sought as output, wrong decisions concerning what to include and what to exclude from such a system are a virtual certainty. In similar fashion, ambiguous or obsolete cataloging systems can hide as much as they reveal. These limitations are likely to be particularly aggravating in the case of user requests related to many of the newer hybrid disciplines.

Another part of the problem might be attributable to the habits or to ignorance on the part of the prospective user. Many scientists and engineers may simply be uncomfortable interfacing with machines for STI acquisition or be unable to express their needs with sufficient clarity to make effective use of such systems.

Information Specialists. It is no accident, therefore, that we witness the rise of the "information specialist" in conjunction with the development of these systems. In some cases, these individuals are employed by the user firm. In others they are employed by the information utility. Regardless of the position they occupy, the basic function of these specialists is to help the prospective user clarify his problems and specify his needs in such a way as to facilitate the efficient production of STI reference materials.

Actually, these information specialists represent a logical extension of the long-standing practice—particularly in the early stages of a research project—of relying upon knowledgeable colleagues or outside authorities for guidance in STI acquisition. Interpersonal communications have few substitutes in a cost-effectiveness sense for these purposes.

To sum up, direction can be said to be concerned with the "why," "what," and "where" of information acquisition. If this process is to be both efficient and effective, there must be agreement within and among the various groups engaged in this activity inside the company on the objectives to be served, the goals to be met, and the constraints to be observed. There must also be a *structure* for the mobilization and allocation of resources and a *system* for the monitoring and control of information acquisition. This is the "how" of information acquisition.

Structure and System

There are vertical and horizontal dimensions to the information acquisition structure of the firm. As these terms are used here, the vertical dimension refers to the assembly, evaluation, and use of STI at successive stages of inquiry within a firm's research & development department. The process is iterative in character and can be compared to the progressive redefinition or reformulation of a problem in a technical sense. The horizontal dimension refers to informational infeeds from other functional units of the same enterprise. These might include R&D units attached to other corporate divisions and marketing research groups located at the corporate or divisional level.

The Vertical Dimension

Research & development activity takes place at a number of levels of inquiry, including basic research, applied research, product or process development, and testing. Beyond the point of full-scale utilization or commercialization, R&D concentrates on applications and improvements research. A given project can commence—or be terminated—at any point.

Different kinds of STI are sought at each stage. Actually, at the inception of a research project, STI is not so much sought as recognized in terms of its technical and/or commercial significance. These are the STI "signals" that initiate a process of investigation. Although serendipity plays an important role at this point it should also be observed, as the cliché puts it, that "chance favors the prepared mind."

Given a basic technical concept such as that embodied in a proposal and authorized for investigation, the search for STI shifts to one of several foci: clarification of the concept, generation of alternative solutions, precedents for anticipated experiments, and assistance in complex technical calculations. "Inside" resources and "outside" expertise may be solicited with the response conveyed in the form of published documents or file information, written or oral advice.

The solution to a technical problem generally requires both internally and externally generated STI. Most of the latter is nonproprietary in character. However, when this is not the case, interfirm agreements (e.g., licensing) may be required to expedite transfer. If vitally needed STI does not exist, and cannot be readily generated internally, the firm may elect to sponsor R&D efforts externally in the quest for a solution.

Internal R&D Transfers of STI. Within a particular R&D department the responsibility for maintianing and/or acquiring different kinds of STI is often divided up among individuals by job function and professional interest. A distinction is often drawn in this respect between scientific and engineering-oriented disciplines because of their relatively unique approaches to technical problem solving. STI transfers between scientists and engineers are often impeded by these differences, which extend to the formats of their respective professional reference materials.

Formal mechanisms for integrating and applying the various kinds of discipline oriented information to a given problem generally involve the formation of temporary project teams composed of appropriate scientific and engineering personnel. Matrix forms of R&D organization are a relatively recent innovation in this respect. Informal mechanisms are exemplified by technological "gatekeepers" who are consulted by their R&D colleagues for advice concerning the sources of specific information.

Within an R&D department, the acquisition and use of STI is complicated by the transfers involved as a particular project moves through its various phases to final completion. Acquisition and use is further complicated by the need to integrate information across technical disciplines at many stages in this process. From the firm's point of view, however, the most serious challenges by far are the barriers encountered in moving information across company divisions and departments.

The Horizontal Dimension

The horizontal dimension of STI acquisitions reflects the fact that other units of an enterprise apart from R&D have a potential contribution to make to new product and process development.

Marketing research, for example, can influence the direction of new product development activity by identifying or clarifying marketplace needs. This is essentially an iterative process. As Table 3.2 reveals, marketing research inputs can help to focus scientific and technological inquiry at virtually all levels of research & development activity.

Consider, for example, the implications of high and rising energy costs for the packaged foods industry. Marketing research could identify and clarify significant commercial opportunities for R&D effort at a number of levels including basic research (e.g., irradiated foods); applied research (e.g., freeze-dried products with superior texture and flavor); and product development (e.g., retortable pouch packaging) to give but three instances.

Similar conclusions could be drawn for horizontal transfers

TABLE 3.2*

PROSPECTIVE MARKETING RESEARCH CONTRIBUTIONS

TO TECHNICAL RESEARCH & DEVELOPMENT

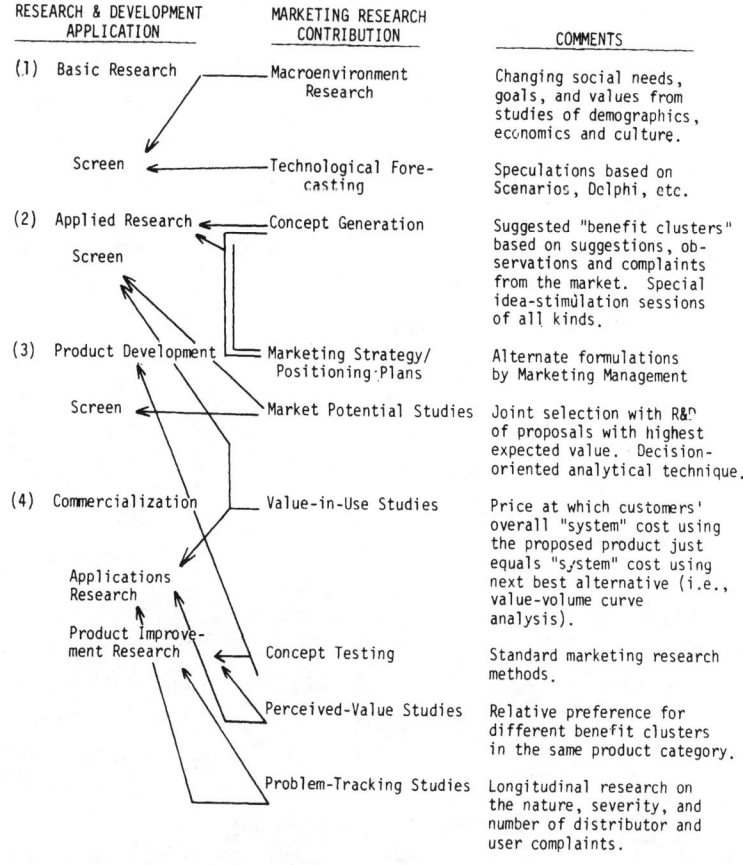

RESEARCH & DEVELOPMENT APPLICATION	MARKETING RESEARCH CONTRIBUTION	COMMENTS
(1) Basic Research	Macroenvironment Research	Changing social needs, goals, and values from studies of demographics, economics and culture.
Screen	Technological Forecasting	Speculations based on Scenarios, Delphi, etc.
(2) Applied Research	Concept Generation	Suggested "benefit clusters" based on suggestions, observations and complaints from the market. Special idea-stimulation sessions of all kinds.
Screen		
(3) Product Development	Marketing Strategy/ Positioning Plans	Alternate formulations by Marketing Management
Screen	Market Potential Studies	Joint selection with R&D of proposals with highest expected value. Decision-oriented analytical technique.
(4) Commercialization	Value-in-Use Studies	Price at which customers' overall "system" cost using the proposed product just equals "system" cost using next best alternative (i.e., value-volume curve analysis).
Applications Research		
Product Improvement Research	Concept Testing	Standard marketing research methods.
	Perceived-Value Studies	Relative preference for different benefit clusters in the same product category.
	Problem-Tracking Studies	Longitudinal research on the nature, severity, and number of distributor and user complaints.

* This table is adapted from a paper dealing with
   Marketing Research/R&D Linkages delivered at Rutgers
   GSBA on March 11, 1977 by Irwin Gross, Manager of
   Marketing Research (Advertising), E.I. duPont deNemours.

of information from the manufacturing function of the firm insofar as new process development work is concerned. In both instances, however, the transfer of relevant event and prescription information may be impeded by organizational rigidities as well as by interpersonal differences.

The formulation of R&D goals, of course, is a shared responsibility involving R&D and other business functions. One cause of problems in information transfer may be that this sharing of responsibility may be restricted to senior management personnel and never properly understood or appreciated at lower echelons of the organization where the work is actually carried out.

A second cause of problems may be the lack of understanding or appreciation of where and how effective interdepartmental coordination may take place. Apart from the initial identification of problems or opportunities, the suggestions of other departments may be viewed as distractive to the realization of narrowly conceived technical objectives.

Interpersonal conflicts are also frequently a cause of horizontal information transfer difficulties, particularly between marketing and R&D. Cooperation and coordination at this interface are often hindered by the basic differences in philosophy, attitudes, and professional goals that distinguish the social scientists engaged in marketing research from the physical scientists and engineers engaged in R&D. All of these problems adversely affect the efficiency and effectiveness of R&D, including its acquisition and use of STI.

Improving the permeability of departmental interfaces is a difficult and challenging undertaking even under the best of circumstances. Decentralized technical research departments within the same company provide a case in point. Here a situation may arise in which the knowledge and skills possessed by one unit are needed by a second in order to realize a particularly attractive market opportunity.

Formal and informal mechanisms are required to identify and locate technical resources within the firm, define their areas of application, and effect transfers when and if this is thought desirable.

The matching of capabilities and needs is the first step. Formal mechanisms used for this purpose include liaison scientists, in-house technical information centers, company-sponsored colloquia, house organs, and interdepartmental committees of all kinds. Informal matching techniques are exemplified by the proverbial company grapevine.

Effecting transfers of STI between these R&D groups is another matter. This involves considerations of status and rivalry as well as risk and reward at the level of both the organizational unit and the individual researcher. At both the unit and the researcher level, the possession of knowledge is the possession of power. It is not given up willingly. Thus, when such transfers do occur it is no accident that they have to be imposed on the parties involved from above. Examples include the transfer of R&D files and personnel from department to department and the reorganization or consolidation of all R&D operations within the company.

Conclusions

Several themes keep reappearing throughout this paper: First, scientific and technological knowledge is accumulating faster than it can be absorbed by the firm in the context of its own needs. Second, STI is sought out by the firm to help it identify potential threats and opportunities as well as to help it solve prespecified technical problems. Third, many of the difficulties the firm encounters in acquiring STI from external sources find their parallel in the obstacles that the firm must overcome in transferring STI internally.

The process of STI acquisition and use is enormously complex. Improving this process is equally challenging. It should be obvious that the enthusiasm and creativity of the individual researcher play pivotal roles at many points in the process. However, these qualities have to be balanced with objectivity if the process of STI acquisition and use is to be both effective and efficient.

Left to their own devices, many researchers are perfectly capable of running amok, as Alec Guiness so beautifully demonstrated in *The Man in the White Suit* some years ago. Overly controlled, their output can consist of a thousand variations on a minor theme, none of which make any significant contribution to the realization of the firm's basic objectives.

What is needed here, then, is disciplined creativity. To foster creativity, the researcher has to know, in general terms, not only what it is the firm is seeking but why, on the chance that he might be able to discover a superior alternative. The researcher must be given incentives not only to seek out relevant information but also to transfer it to others within the firm on the basis of anticipated or existing need. Finally, the researcher must have the freedom to search out relevant STI internally or externally with as few restrictions on his depth or range of inquiry as possible.

Discipline is also essential. Thus, research incentives to seek out STI must be balanced with disincentives for inappropriate disclosure of confidential information to outsiders on one hand, and for failing to cooperate freely and fully with others inside the organization in their own quest fo STI, on the other. Similarly, research freedom of access must be balanced with the need to periodically justify the time and money expended on STI search. Training in the art of STI acquisition and use would also seem to be needed to ensure that this freedom of access is utilized productively.

If the firm is to take advantage of the unparalleled opportunities represented by the explosion of technical knowledge, it must improve its acquisition and use of STI. These improvements will undoubtedly take the form of changes in the direction, structure, and system of these information flows. In a larger sense, however, whether these changes will actually constitute improvements in the flow of relevant technical information will depend ultimately upon the innovational climate they engender within the firm at the level of the individual researcher.

4

A Model of User Behavior for Scientific and Technical Information

Jagdish N. Sheth

Introduction

There are a number of compelling reasons why scientific and technical information (STI) systems should become more user oriented in terms of content and design of the information, the process of dissemination, the degree of availability to potential users, and the pricing policies that the system uses.

First, as is commonly the case with resources such as information, less than 20% of all potential users probably make 80 to 90% of the actual usage of STI. There is some indirect evidence of this tremendous skewness in usage of STI from the fact that many scholarly journals are subscribed to or read by only a small percentage of their total potential users. In fact, this is often the reason why STI documentation in the form of journals, books, pamphlets, and monographs incur losses and have to be subsidized by outside sources such as grant-giving agencies, foundations, professional associations, and academic institutions.

Second, even among those who regularly use STI, it is not uncommon to find a sense of frustration and dissatisfaction with the existing documentation with regard to content, format and accessibility. For example, documentation may not be available at a time and place that is convenient to the user. This clearly indicates a dissemination-distribution problem. Furthermore, when a potential user has access to STI documentation, he may find that it is hard to comprehend. Even if he can comprehend the content of an STI documentation, it may be packaged in a format that makes it difficult to use on a continuing basis. Finally, many users of STI find it difficult to separate the small quantity of relevant information from the large amount of the irrelevant that is conveyed to them.

Greater user orientation in the design and dissemination of STI is inevitable as the technology of STI becomes mature. As a product that is entering a more mature phase in its life cycle, th STI system can no longer afford to continue to be producer oriented. While it is perfectly acceptable and sometimes even necessary to put greater emphasis on the production side in the beginning of a product's development, every product and service must become more user oriented as it gains acceptance and establishes a base among its users. Otherwise, it is likely to experience a crisis of relevance and usefulness. Such a crisis is currently alleged by some STI users.

Benefits of User-oriented STI System

A number of benefits are obvious in a user-oriented STI system.

The first, and probably the most important, benefit is to extend the product life cycle of STI by searching for new segments of users such as foreign scholars and researchers as well as by disseminating the information to nontechnical but educated people in the society by satisfying their epistemic needs. In fact, with small changes in the content and format of STI, it may be possible to create outreach programs far beyond the boundaries of immediate users. This is already manifested by the existence and success of many magazines like the *Science Digest, Psychology Today, Scientific American,* and to some extent *Harvard Business Review.* In short, a user-oriented STI system can easily reveal at least three to four levels of interest in scientific and technical information.

Second, a user-oriented STI system is also likely to reveal the true nature of complementary and competitive relationships among various types of STI documentation. For example, we really do not know whether the "annual review" is perceived as a substitute or complementary reading to many of the journals in psychology, sociology, and anthropology. Similarly, many fear that the journals may have become an obsolete source of receiving bibliographic information with the advent of computerized search procedures that are available today in many fields.

Third, a user-oriented STI system may force the innovation of new concepts, ideas, formats, and products. Many argue that there is a serious problem in the area of scientific and technical books, where the costs have gone ever upward and markets have become miniscule. It is not surprising that most of the current efforts are in the direction of cost-reduction tactics such as photocopy publishing and soft-binding, instead of a more fundamental change in the direction of producing and disseminating scientific and technical information in nonbook formats. Such a fundamental change will occur when publishers of STI understand and utilize the psychology of the users.

Fourth, any STI system that is user oriented is likely to increase its frequency of use. By definition, such a system is designed, produced, and disseminated in a manner that increases its likelihood of being used more often by the potential users. For example, a number of publishers of STI have realized the need for greater illustrative and pictorial materials that facilitate comprehension on the part of the reader. Similarly, some libraries arrange the publications in the open stacks in a manner that facilitates greater use. Unfortunately, these are still not very common practices, but rather exceptions.

Finally, a user-oriented STI system is likely to survive any crises of social relevance. It is not at all difficult for people to become skeptical toward scientific and technical information that remains highly abstract and portrays a picture of elitist attitudes. In the long run, any STI cannot survive if public opinion is negative toward it. We have already witnessed some examples of negative public opinion in the form of "fleece of the month" awards given by Senator William Proxmire in the United States.

Modeling STI User Behavior

In order to make STI system more user oriented, we must first understand and model STI user behavior and its underlying psychologyical processes of perceptions and motivations. Fortunately, there is a good deal of knowledge and research in the area of consumer behavior that seems relevant toward modeling STI user behavior (Howard and Sheth, 1969; Engel, Kollat & Blackwell, 1973; Hansen, 1972; Sheth, 1974; Ward and Robertson, 1974).

Based on a model of individual choice behavior (Sheth, 1975), an attempt is made in this paper to model the perceptions, values, and usage of STI. It is summarized in Figure 4.1. There are two major and distinct areas of conceptualization in the model. The first part relates to perceived STI utility in the minds of potential users, and the second part relates to actual usage behavior of STI system. These two areas are represented as *A* and *B* in Figure 4.1 and are described in detail in subsequent sections.

A. Perceived STI Utility

The utility of STI among potential users is considered to be a function of the potential user's perception of the extent to which STI satisfies a set of needs, wants, and desires. In other words, each potential user has a set of evaluative beliefs about the usefulness of STI related to different types of needs. We have identified five different types of needs on which STI may be evaluated. They are functional, problematic, social, emotional, and epistemic needs. These are shown in Figure 4.1 to influence the perceived STI utility.

Depending upon how closely the evaluations of a particular STI system match with the expectations of the user in terms of these needs, each STI system is perceived to possess some degree of positive or negative utility which becomes the basis for its eventual usage. The greater the matching, the higher the perceived utility of a particular STI system, and vice versa. Mathematically, this can be expressed as follows:

$$U\,(\text{STI})_k = \sqrt{\sum_{j=1}^{5} \left(E_{x_j} - E_{v_{jk}} \right)^2}$$

The utility model presumes that a particular STI system can acquire negative utility by offering too little or too much of a particular need satisfaction. Furthermore, it is quite possible

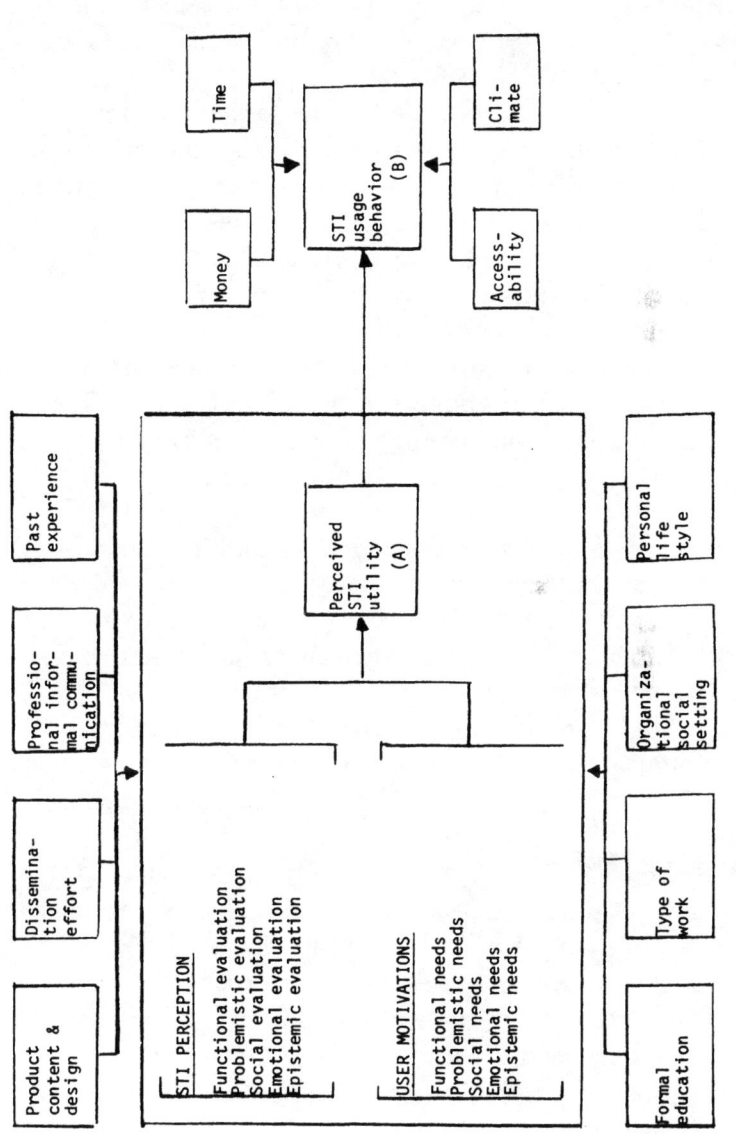

FIGURE 4.1: A MODEL OF STI USER BEHAVIOR

that the perceptions of a particular STI system in the minds of potential users may not necessarily match with its objective reality due to either stereotyping or lack of full information.

We will briefly describe the five types of needs which can be satisfied by usage of a STI system.

1. *Functional needs* are generated by task-related activities. For example, a full time research scholar in the academic setting is often promoted exclusively on the basis of his record of scholarly publications and research. In the process of producing scholarly research, he needs scientific and technical information. Similarly, a professional working on applied areas also has functional needs for scientific and technical information. However, his functional needs in terms of both content and format may be very different from those of the research scholar. Therefore, a particular STI system may be perceived as more useful to one person and less useful to the other person.

2. A second type is the *problematic need* for STI. In this case the need for STI is not absolute but conditional upon a set of situational contingencies or antecedents. For example, the need for certain STI becomes manifest among graduate students when they are assigned the task of writing a research paper on a certain topic. In fact, most textbooks are often regarded as possessing only problematic utility since they are relevant, at least in terms of student perceptions, primarily for a particular course. This goes for all the reading assignments in the course, whether they are in a periodical, book, monograph or some other medium. The peak demand experienced by libraries for the assigned reading materials is clear evidence of the problematic utility they possess from the user's viewpoint.

3. The third type of utility for a particular STI system is created not because of its intrinsic value, but because of its association with certain social roles and stereotypes. This is referred to as the *social needs* of STI. For example, many professionals in their offices and homes display certain books and periodicals primarily to reveal their organizational role identification. In short, many STI books and periodicals are needed for conspicuous consumption purposes. This is especially true for handbooks and professional encyclopedias.

4. A fourth type of STI utility is based upon its satisfaction

of personal *emotional needs*. Once again, the particular STI system has less functional or intrinsic value to the individual. Instead, it acquires more of an emotional or extrinsic value because of its association with a particular discipline, author, or publisher. For example, many people possess complete writings of a particular author because they pride themselves as collectors. Similarly, some people use a particular STI product such as a book or a periodical because they are emotionally attached to it by prior learning or conditioning in their earlier days as students and research assistants. This seems to be particularly true for more technical subject areas such as statistical and mathematical books. It is even suggested that some books remain highly popular as textbooks in a discipline due more to the emotional attachment on the part of the instructor than to their intrinsic superiority over other books for the course.

5. The last type of need satisfied by STI is the *epistemic needs* of the potential users. Epistemic needs refer to the human desire to be knowledgeable, inquisitive, and curious about phenomena which surround them but are not directly related to their job activities. They also represent acquisition of information now which may have some potential functional utility sometime later in life. We think that epistemic needs dominate in the determination of perceived utility of STI, and probably rank in importance only next to the functional needs.

In summary, the perceived utility of STI is presumed to be a vector of five distinct dimensions reflecting the degree to which it satisfies the functional, problemistic, social, emotional, and epistemic needs of potential users of scientific and technical information.

B. Individual and Product Differences

Unfortunately, perceived STI utility is subject to both individual and product differences. In other words, it is likely that different types of scientific and technical information, documents, areas, and disciplines will have different vectors of the peceived utility among potential users. Similarly, a particular type of STI will have a distribution matrix of perceived utility among a sample of potential users since both the needs as well as perceptions are likely to vary from individual to individual.

It becomes, therefore, essential to theorize about some of the major determinants of individual and product differences. We have isolated four individual-related factors and four product-related factors that seem to be most relevant and useful for modeling the differences between users and between types of STI documentations. These four functions are described in Figure 4.1

1. *Individual Factors.* The individual user-related factors are:

a. Prior education,
b. Type of work,
c. Organizational-social setting, and
d. Personal life style.

It is obvious that the level and type of *prior education* will often determine the specific types of need satisfaction in an individual user of scientific and technical information. Prior education will determine both the attitude or value system toward STI in general as well as shape the degree of expectations with respect to the functional, social, problematic, emotional and epistemic needs. For example, an individual with less than a high school education is likely to have less involvement and interest in STI than another individual with a doctorate degree. Similarly, a doctorate in psychology will produce a different vector of expectations than a doctorate in medicine, for example. Even though prior education seems such an obvious factor in determining individual needs for STI, it is surprising that there is very little research on the topic.

A second individual-related factor is *type of work.* Once again, it seems obvious that there should be strong differences between blue collar and white collar workers as well as between clerical and professional white collar workers with regard to the level and type of needs for STI. However, it is possible to detect even more subtle individual differences among the white collar professionals who are likely to be the prime target market for STI by examining the typology of work in terms of staff vs. line functions, academic vs. professional endeavors, and research vs. administrative responsibilities.

The third individual-related factor is the *organizational-social setting.* It is argued that the social structure and organizational structure surrounding an individual user will also determine and influence whether he will have need for scientific and technical information and if so, whether it will be more a functional, epistemic, or social need for STI. If consumer behavior is any guide in this area, one would presume that organizational-social setting will primarily generate the social (conspicuous consumption) and epistemic needs for STI. This is because certain organizations and social classes tend to acquire images or stereotypes in which it is fashionable to possess, if not use, scientific and technical information.

For example, many research laboratories affiliated with large corporations pride themselves on maintaining excellent libraries on STI which may not be relevant to the task or utilized by the research employees. Similarly, the upper socioeconomic classes may take pride in possessing, if not reading, nonfiction literature including scientific and technical information. In fact, in many upper socioeconomic classes the possession of STI documentation is conspicuously displayed for the outside world as if to indicate that mass media such as television, radio and popular magazines do not serve to satisfy their epistemic and social needs.

An individual user tends to be influenced by the organizational-social setting surrounding him in one of two ways. First, such a setting may influence whether he should or should not be a user of scientific and technical information. Second, a particular type of STI may or may not be considered as appropriate for him to use. For example, it is this type of influence which often creates the dichotomous world of trade journals and scientific journals or of trade books vs. academic books.

The last individual-related factor is the *personal life style* of the individual user. We believe that scientific and technical information may be perceived as having or not having a useful role in the personal life style of an individual as indicated by his daily activities, interests, values, and opinions. Personal life style is likely to generate the emotional and epistemic needs for STI. For example, an individual aspiring to an intellectual as opposed to a "swinging" or family-oriented life style will certainly develop a need for STI. It seems that there is a sufficient degree of empirical research on personal life styles (Wells, 1974) to develop a typology of potential users of STI and hypothesize the distinct roles scientific and technical information play in the daily living patterns of people.

2. *Product Factors.* There are four product-related factors that produce differences in the perception of STI utility from one type of STI to another, such as between books and journals, between hard sciences and social sciences, and between trade and academic publications. Thcsc factors, as shown in Figure 4.1, are:

a. Product content and design,
b. Dissemination efforts,
c. Prior familiarity and experience, and
d. Professional informal communciation.

The product itself, in terms of its *content and design*, is likely to be the single most important factor determining differences between various forms of STI. For example, encyclopedias and handbooks tend to be useful in a different way than journals in satisfying any of the five needs discussed earlier. Similarly, abstract periodicals are perceived somewhat differently than regular journals. While it is obvious that varying content of STI documentation will certainly result in interproduct differences, we think there has been inadequate emphasis on the design aspect of the products. This includes format, writing style, medium of representation such as language vs. pictures, and packaging aspects. In this regard, STI producers can learn a great deal from advertising agencies and commercial publication houses.

Dissemination effort is the second product-related factor. It includes the conscious allocation of resources in informing, communicating, and influencing the potential readers about the availability of the STI system, and in making the product as easily accessible to the potential user as possible. In marketing, this would imply allocation of resources to distribution and promotion efforts by which time and place utility are added to the product. In the STI area, the university presses and publica-tion bureaus are generally backward in this regard as compared to commercial publishers. We think that STI can learn a great deal in the area of dissemination effort from the marketing area.

A third product-related factor is the individual user's *past experiences* with the STI system in general, and specific to a particular STI product under question. Users learn a great deal by trial and error about various STI products, and their evaluations will be significantly shaped by the degree of satisfaction they experience with a particular STI product. It is not at all uncommon for many STI products to have a large percentage of transient users who use the product irregularly as well as switch around from product to product, probably in hopes of finding one or two ideal products for their needs.

The fourth product-related factor is the *professional-informal communication* specific to a particular STI product or system. The influence of word-of-mouth communication such as reviews, comments, references, citations, etc., seems to be enormous. Many scholars rely on others to separate good STI products from poor ones either because they cannot cope with the flow of indiscriminating dissemination of STI or because they feel a sense of risk in adapting unfamiliar STI products. In fact, in many social science disciplines, we see the emergence of review periodicals that seem to cater to these people. *Contemporary Psychology* is a good example of this type of professional informal source of communication and influence in psychology. Similarly, the popular textbooks in every area also perform the same function. Finally, many individuals acting as professors, consultants, and advisors to others perform a comparable function of gatekeeping and opinion leadership with regard to a particular source of scientific and technical information.

Between the individual-related and product-related factors, it may be hypothesized that the former factors primarily shape and change the user's needs and expectations, while the latter factors shape and change the user's perceptions and evaluations of specific STI products, sources, and systems. In any event, the model clearly suggests that we need to adapt a market segmentation strategy for adequate dissemination and usage of scien-

tific and technical information. There is no way a universal STI
system can be designed with which we can satisfy all the potential users. In short, STI cannot be all things to all users.

3. *STI Usage Behavior.* A second part of the model of STI
user behavior relates to the actual usage of STI products,
sources, and systems. There are at least three dimensions of
STI usage behavior that need to be fully understood and
analyzed.

1. The first dimension is the *selectivity* of specific STI
   products, sources, and systems a particular individual user
   makes use of. By a microlevel individual user analysis, it
   is possible to assure the degree of selectivity of usage of a
   particular STI product, source, or system. It should then
   be possible to identify which specific STI products,
   sources, and systems are used by the same users indicating
   a measure of complementarity among them.
2. The second dimension is *the amount of usage* of a particular STI product, source, and system. It is argued that there
   will be a great deal of variability in the usage rate of a
   particular STI product within the user segment so that
   some will be heavy users and others will be light users. By
   analyzing the heavy half users, it is possible to measure the
   degree of skewness in usage and, therefore, the aforemen-
   tioned 20/80 ratio: less than twenty percent of total
   potential users often generate more than eighty percent of
   total usage. We believe that the skewness is even greater in
   the case of certain STI products and therefore, it might be
   worth the effort to assess the cost/benefit ratios for various STI products and services.
3. The third dimension of STI usage is related to *user loyalty*.
   Does a user acquire a habit toward a particular set of STI
   products, sources, and systems by experience and learning,
   and, if so, does this habit become a deterrent for the diffusion of new and innovative STI products? Loyalty represents continuous use over a fairly long time period and
   without strong interruptions for a given STI product. For
   example, in the case of college textbooks, it is this product
   loyalty which often is the basis for a virtual monopoly by

a particular author. Usage loyalty is indicated by the systematic repeat choices a user makes in favor of a particular STI product even though he has the opportunity to switch to others. For example, the relative frequency of renewal of journal subscriptions would be one indicator of user loyalty.

The model presumes that the perceived STI utility of a specific product, source, or system is only a necessary condition for its actual usage. There are at least four major sufficient conditions: time, money, accessibility, and climate. We believe that often a good STI product with a high degree of perceived utility remains underutilized by the potential users because of time, money, accessibility, and climatic factors.

Increasingly, time is becoming a bigger constraint as compared to money requiring many STI products and systems to adapt different formats and packaging devices to remain viable for usage under extreme time constraints. This is especially true with the advent of computerized storage and retrieval systems designed for bibliographic research, for example. However, very little systematic research is made as yet to manage the other two sufficient conditions, namely accessibility and climate. It seems that accessibility is a serious weakness in many existing STI systems and products especially under peak pressures. One hopes that computer technology will be very useful in this area as it becomes increasingly applied in the distribution of scientific and technical information. For example, the current library procedures of binding journals, creating a serious accessibility problem for periods as long as eight to twelve weeks, could be replaced by alternative technologies.

We also believe that climate will also receive increasing attention in the near future. In the sense used here, "climate" refers to the physical environment in which STI is to be used and thereby its influence in inhibiting or enhancing the usage behavior. For example, pleasant atmospheres are generally more conducive to product usage. The atmospherics can be related to the institutional environment, personal office environment, or the laboratory environment. It is a sadly neglected area so far in the area of scientific and technical information.

Linking User Behavior for Planning
and Marketing of STI

Understanding and modeling the user behavior itself is not relevant or useful. It must be systematically incorporated in the planning and management of STI. In this section, we have attempted to integrate the STI user psychology and behavior in the total process of planning, producing, and disseminating scientific and technical information. The integration is represented in Figure 4.2 and is based upon a similar effort in the area of planning of multinational corporations (Sheth, 1977).

We very strongly believe that the STI planning process should be based on identification and continuous monitoring of changing user needs. The monitoring itself can be done by a separate research unit called "basic research." There should be an ongoing interaction between STI planning and basic research in view of the fact that there may be technological breakthroughs in the supply function. In other words, the role of the STI planning unit is to bridge the fundamental supply and demand factors.

The STI planning unit should be given the task of developing new concepts, products, and services by blending together the user needs on the one hand, and legal, technological, and financial considerations on the other hand. It is possible that some of the user needs may not be fully recognized in the development of new concepts because of technological, legal, or financial considerations. However, it is the task of the STI planning unit to seek optimization between user needs and technological-financial factors.

The new concepts developed by the STI planning unit must then be brought to reality and managed—the task of STI management. It consists of putting flesh on the skeleton by producing and distributing a set of STI products, systems, and services. This will require integration of production economics, distribution channels, and marketing efforts. The latter include the pricing and promotion aspects.

The result of STI management function will be the dissemination effort. However, we strongly argue for a *differential* dissemination effort by properly targeting the production,

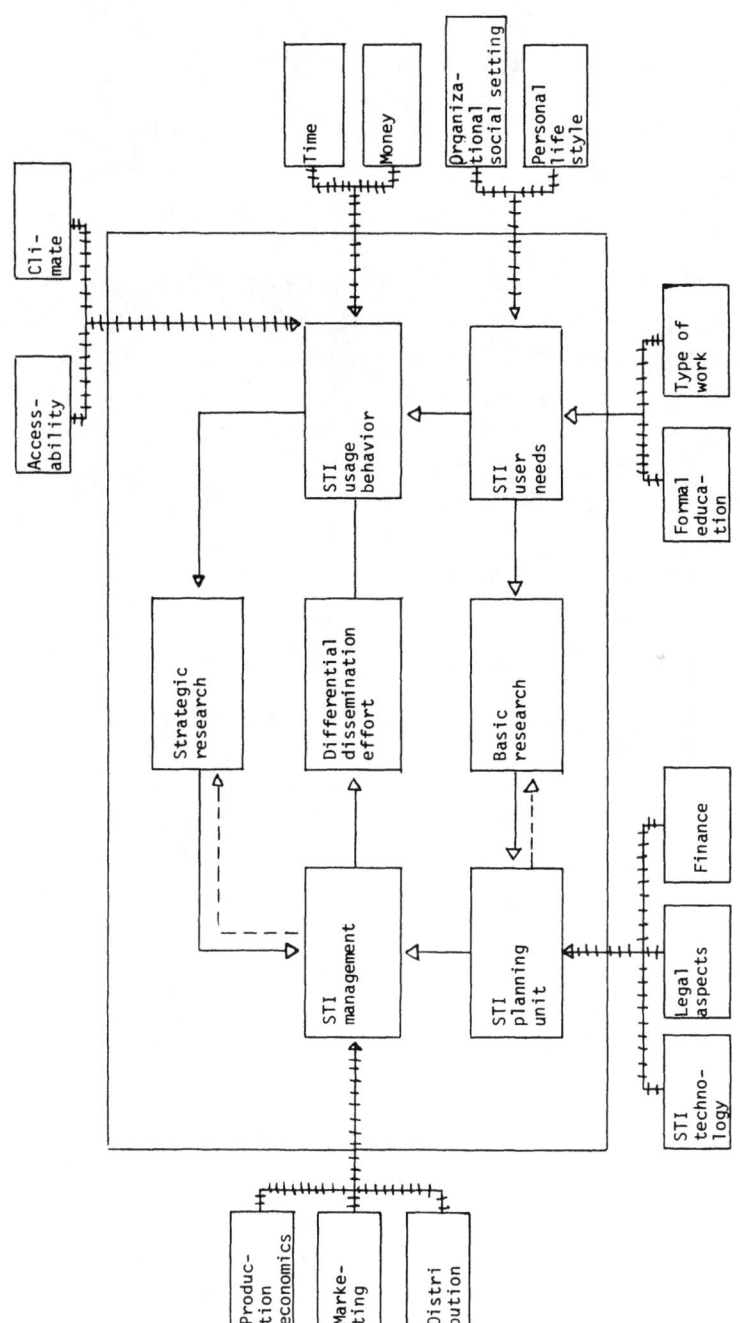

FIGURE 4.2: PLANNING AND MANAGEMENT OF THE STI FUNCTION

marketing, and distribution activities to selective user segments. As mentioned before, there are simply too many individual user differences in reality to hope for a universal product, distribution, and marketing policy. The differential dissemination effort combined with user needs will result in perceived STI utility. This will in turn manifest itself in STI usage behavior depending upon the time, money, accessibility, and climate conditions.

We also suggest that the usage behavior and the underlying necessary and sufficient conditions should be monitored by STI management. This type of monitoring activity is called "strategic research" in Figure 4.2. Hopefully, STI management will have the insights to design and implement dissemination efforts in such a way that the feedback obtained through strategic research in both pretesting and post-implementation testing will be in the direction of desired objectives such as achieving projected levels of STI usage in various targeted segments.

Summary and Conclusions

It is our hope that this paper has shown the relevance of consumer behavior to the task of disseminating scientific and technical information. The basic theme of the paper has been that to produce and distribute scientific and technical information without proper inputs of the user needs and psychology is at best a wasteful and highly myopic process. As such, it cannot survive too long in a society that believes in free and voluntary choice open to the users of scientific and technical information.

References

Engel, J.; D. Kollat; and E. Blackwell. *Consumer Behavior.* Holt, Rinehart, 1973.

Hansen, F. *Consumer Choice Behavior.* The Free Press, 1972.

Howard J., and J. Sheth. *The Theory of Buyer Behavior.* Wiley, 1969.

Sheth, J. N., ed. *Models of Buyer Behavior.* Harper & Row, 1974.

Sheth, J. N. *Toward a Model of Choice Behavior.* Paper presented at the AMA Doctoral Consortium, Cornell University, 1975.

Sheth, J. N. "A Market-Oriented Strategy of Long-Range Planning for Multi-national Corporations." *European Research* 6 (January 1977):3-12.

Ward, S., and T. Robertson, eds. *Consumer Behavior.* Holt, Rinehart, 1974.

Wells, W. D., ed. *Life Styles and Psychographics.* American Marketing Association, 1974.

5

Problems and Prospects in the Segmentation of the STI Market

Yoram Wind
Robert Thomas

Introduction

The relevance and applicability of marketing approaches to the design and dissemination of STI systems is hardly debatable. Our presence here—at a conference devoted to the marketing of STI—and the continuous National Science Foundation support of research in this area are strong indicators of the inroads marketing has made in this important industrial area.

Market segmentation, as one of the fundamental and most applied theories of the marketing discipline, is thus widely accepted conceptually as one of the key building blocks of any marketing strategy for STI systems. Yet, in practice, most of the STI marketing efforts have ignored the segmentation concept.

This gap between the conceptual acceptance and practical rejection of the segmentation concept can in part be due to the difficulties involved in *implementing* a segmentation research program and utilizing the results in the design of STI marketing strategies. The objectives of this paper, therefore, are to briefly review some of these difficulties and suggest some possible solutions. The discussion is organized around the five major research phases: (1) segmentation problem definition, (2) research design, (3) data collection, (4) data analysis, and (5) data interpretation and "translation" of results.

Segmentation Problem Definition

The segmentation problem definition stage is probably the most crucial and most neglected area of segmentation. This is the stage in which management should ask the question, "Why segment the market?" In attempting to explicate an answer to this question, management and their researchers should develop a *segmentation model* that specifies the desired bases for segmentation as well as the specification of the desired descriptor variables. The selection of specific bases for segmentation depends on the way segmentation results are to be utilized. If, for example, management concern is with increased usage of STI service by current users, a natural basis for segmentation would be the usage status of organizations in the relevant market. (Note, however, that by restricting analysis to a "relevant market" management is, in fact, adding another level of segmentation. For example, if the relevant market is defined as High Intensity R&D Industries, this serves as an initial basis for segmentation—R&D intensive SICs [standard industrial classification] vs. other SICs.) In addition, possible descriptor variables have to be specified. These variables should include those that can help identify and reach the desired segments (Wind and Lotshaw, 1973).

The specification of a segmentation model is not a trivial task and requires the resolution of a number of major questions, including:

a. What should the unit of analysis be? Should it be the organization, individual respondent within the organization, or the buying center?

b. Can one assume intraorganization homogeneity (the commonly made assumption)? And if this assumption seems questionable (as suggested by some recent research for some products and buying situations), how should the intraorganization heterogeneity be identified and handled?

c. How stable are market segments? (How often do firms—individuals or buying centers—move from one market segment to another?) And should one undertake a large scale base line segmentation study or apply the segmentation concept on an ongoing basis to all the firm's research activities?

d. How "flexible" should the segments be? I.e., can the composition of segments change easily or does an organization (or individual), once included as a member of a segment, remain with this segment? Typical segmentation approaches have implicitly assumed a fairly permanent membership. Yet, recent advances in the utilization of computer simulation in segmentation research enables the researcher to develop "flexible" segments. These flexible segments are based on the respondent's specific reaction to various marketing offerings (as measured, for example, via a conjoint measurement procedure [Green and Wind, 1975] and incorporated in a computer simulation, which results in a share of choices for each offering given alternative competitive strategies. For a discussion of this procedure, see Wind, Jolly and O'Conner [1975].)

Typical answers to these questions by the current segmentation studies of the STI market are summarized in the left column of Table 5.1. In contrast, the right hand column presents our belief as to the direction in which we should move in STI (and other organizational) segmentation studies.

Research Design

Market segmentation studies require research designs that are responsive to the requirements of the segmentation model. The more thorough the segmentation model the higher the likelihood that "standardized" research procedures will not be appropriate and the more creative and imaginative research design will be called for. Consider, for example, the research design implication of using the buying center as the unit of analysis and assuming intraorganization heterogeneity.

Some of the specific decisions involved in this stage include:

a. Conceptual and operational definitions of all variables (including specification of type of scales used—nominal, ordinal, or interval).
b. Determination of whether to employ a longitudinal or cross sectional design.
c. The selection of a laboratory or real-world study.
d. Sample selection.
e. Determination of respondents' task.
f. Selection of the analytical procedures to be employed.

Too often, these decisions are made following the format used in some earlier studies. Although comparability with previous studies is important, it should not replace a systematic

TABLE 5.1

"CURRENT" SEGMENTATION MODELS IN THE
STI MARKET VS. AN "IDEAL" MODEL

	"TYPICAL" CURRENT MODELS	
BASES FOR SEGMENTATION	• Org. Demographic • Product Usage • Needs	Cannot be determined in advance since it depends on specific management needs. Consideration should be given, however, to a 2-step approach to segmentation [Wind and Cardozo, 1974] and cri-teria used in purchase de-cisions [Wind, 1973].
UNIT OF ANALYSIS	• Organization • Individual	• Buying Center [Wind, 1977]
INTRAORG. HOMOGENEITY	• Assumed	• Tested for and proce-dures for dealing with intraorganization heterogeneity developed [Wind, 1976]
BASE LINE VS. ONGOING SEGMENTATION	• Base Line Studies	• Ongoing Segmentation Studies
DEGREE OF FLEXIBILITY	None	"Flexible"

and comprehensive evaluation of alternative courses of action. Consider, for example, the simple variable "STI usage"—how should it be defined? Should it be defined in terms of number of times an STI system is used over some time period? Should it be the cost involved? The number of searches conducted? The number of references provided? The number of times each individual used the system? The number of users? The time lapse between uses? The share of the STI system of all scientific and technical information searched by a given individual or organization?

Furthermore, how should such data be collected—by questioning the user or the librarian? By keeping records of all actual searches? Or by some other unobtrusive measures?

Unfortunately most STI segmentation studies have given little attention to these and similar research design questions.

Data Collection

To date, most segmentation studies of STI systems have relied on primary data collection. Yet, the accumulation of various STI data banks might suggest the possibility of greater reliance on secondary sources.

No specific data collection procedure—personal, telephone, or mail, nor the use of pencil and paper, interpersonal, or computerized interactive interviewing procedure—can or should be singled out as the most appropriate. The selection of a specific data collection procedure should reflect the requirements of the research design, the various biases that might be associated with it, and management's time and monetary constraints.

As with research design, most STI segmentation studies have tended to be too conventional (and not very creative) in their data collection procedures. A number of attempts have been recently undertaken at exploring the feasibility of new approaches to the collection of STI data (Wind and Myers, 1977). Further work along these lines is required.

Data Analysis

Conventional segmentation research for consumer and industrial products and services, as well as for STI systems, has focused on two distinct analytical steps—the determination of the number of segments (either on an a priori basis or in post hoc type studies based on the results of some cluster analysis), followed by the establishment of the segments' profiles (using procedures such as multiple discriminant analysis, multiple regression analysis, and the like).

Despite the wide attention given to and usage of a priori and post hoc segmentation procedures, there seem to be a number of unresolved conceptual and methodological issues which could affect the actual utilization of segmentation in the STI market. Consider, for example, the following problems (which affect not only the data analysis phase but the entire research design and all subsequent research phases):

a. *Intrasegment heterogeneity.* In a priori segmentation, one often finds that the segments *are* different in terms of their mean profiles. This type of data does not reveal, however, the presence of possible subsegments *within* the a priori segments. Consider, for example, the use of subscription status as a basis for segmentation. Members of a subscriber segment may subscribe to the STI service for different reasons; they may be quite heterogeneous in their background characteristics and information needs. Most a priori segments can be decomposed into subsegments or latent classes.

Latent class analysis (Lazarsfeld, 1950; Myers and Nicosia, 1968) can be useful in describing subsegments in terms of: How many are there? What is their relative size? And what are their background characteristics?

Recently Green, Carmone, and Wachspress (1976) proposed a multivariate model called SPA (Segment Partition Analysis), which combines features of latent class analysis and orthogonal array designs. The technique can be applied to any multiway contingency table (of reasonable dimensionality), revealing the extent of heterogeneity in a set of categorical data and the latent classes that make up the total group. Despite the conceptual attractiveness of this procedure, it has not yet been utilized in the segmentation of STI markets.

b. *Determining the number of target segments.* Cluster analysis of benefits, needs, or any other attitudinal or behavioral data results in the segmentation of a market into a number of segments. Statistically, the larger the number of segments the higher the homogeneity of the segments. Yet, from a managerial point of view, there is considerable advantage (in terms of costs and manageability) in selecting only a few target segments. The conditions under which different target market segments can be selected, for a given product and product line, are not well specified, nor are there clear-cut "rules" for the determination of the "best" number of segments.

c. *Comparability of results of various bases and methods for segmentation.* Segmentation research projects vary widely with respect to both the bases they use for segmentation (benefits sought, needs, attitudes, etc.) and the research method they employ (Frank, Massy, and Wind, 1972). Most studies employ one or at most a few bases for segmentation and rely predominantly on a single research approach that reflects the researcher's preferences.

Seldom, if ever, are the results of a number of alternative bases for segmentation and alternative research approaches compared. Yet, if the results are found to be stable (across methods and bases), it would increase the researchers' and users' confidence in, and encourage the implementation of, the results of the segmentation study.

d. *Stability of results.* The question of segment stability—whether members (individuals or organizations) of a given segment remain in the same segment over time—is often a major deterrent to the utilization of the results of segmentation studies. Since no theoretical guidance can be provided, management has to resort to empirical testing for segment stability by continuously monitoring the market. A longitudinal design would enable the assessment of the nature and degree of mobility among segments, and the conditions under which segment stability is likely to prevail.

Componential Segmentation:
An Alternative Approach to Segmentation

Componential segmentation (Green, Carroll, and Carmone, 1975; and Green, 1977) represents a different philosophical and modeling approach to the study of market segments. Unlike a priori or post hoc methods, componential segmentation is concerned with the attributes that underlie the segments, rather than with the specific segments themselves. In principle, componential segmentation can make predictions of how a segment composed of more basic components would react to a test stimulus, such as a new product or service also composed of more basic components.

In a priori or post hoc segmentation, interest is focused on a *fixed set* of specific segments. In componential segmentation, interest centers on the *components* of these segments. In componential segmentation, parameter values are developed for both background characteristics (demographics, product usage, benefits sought, and so on) and stimuli (such as structural or functional properties of STI products). These parameter values are estimated from data obtained from a limited number of selected respondents' evaluations of a limited number of designed stimuli.

Componential segmentation is still a new idea and only a few applications have been carried out. Yet, based on the limited evidence to date, it would appear that it represents a most efficient data collection procedure if one is severely limited in terms of a sample size—a situation typical of most STI studies.

In addition, it provides rigorous insight into the most desirable product characteristics, as developed from conjoint scaling, for each customer segment. The application of componential segmentation to the STI market seems, therefore, to be a natural extension of earlier segmentation studies that utilized conjoint measurement for the determination of respondents' utilities for various features of STI systems (Wind, Grashof, and Goldhar, 1975).

Data Interpretation

The situation in which segmentation based marketing recommendations are ignored by management is not an uncommon one. To reduce the likelihood of ignoring the results of segmentation studies, it is essential that the interpretation stage—"translation" of results into action—be conducted jointly by the researchers and the relevant management team.

This stage requires the ability to translate a segment profile into guidelines for marketing strategy. Consideration should be given here to:

    a. *Product line* (vs. single product strategies), i.e., development of a product line in which each product (or a number of products) is designed (and positioned) for different segments (Wind, Grashof, and Goldhar, 1975).

    b. *"Self selection"* (vs. controlled) communication strategies (Frank, Massy, and Wind, 1972).

    c. Strategies for segments of current STI users *and* segments of nonusers.

Concluding Remarks

Given that the STI market is heterogeneous with respect to needs, perceptions, preferences, usage, and dissemination patterns, the concepts and techniques of market segmentation can, and should, play a significant role in increasing the relevance and effectiveness of information dissemination systems. Better understanding of the segmentation research process and resolution of the conceptual and methodological problems raised in this paper (and others) will hopefully lead to the increased practical utilization of segmentation research in the STI market. The market for scientific and technical information *should* be segmented, target market(s) selected, information dissemination systems designed to meet the needs of the selected target segments, and a marketing program developed to best reach the target market(s) (Frank, Massy, and Wind, 1972; Wind, Grashof, and Goldhar, 1975). Such efforts should not be limited, however, to the traditional segmentation research efforts. The problems associated with these approaches should be identified and new approaches examined and tested.

References

Frank, R. E.; W. F. Massy; and Y. Wind. *Market Segmentation.* Englewood Cliffs, NJ: Prentice-Hall, 1972.

Green, P. E. "Design Considerations in Attitude Research." In Y. Wind and M. Greenberg, eds., *Moving Ahead with Attitude Research.* Chicago: AMA, 1977.

Green, P. E.; J. D. Carroll; and F. J. Carmone. "A Componential Approach to Market Segmentation." University of Pennsylvania Working Paper, November 1975.

Green, P. E., and Y. Wind. "New Ways to Measure Consumers' Judgments." *Harvard Business Review* (July-August 1975): 107-111.

Lazarsfeld, P. F. "The Logical and Mathematical Foundation of Latent Structure Analysis." In S. A. Stouffer, et al., eds., *Measurement and Prediction.* Princeton: Princeton University Press, 1950, chap. 10.

Myers, J. G., and F. M. Nicosia. "New Empirical Directions in Market Segmentation: Latent Structure Models." In R. Mayer, ed., *Changing Marketing Systems*, pp. 247-252. Chicago: AMA, 1968.

Wind, Y. "Recent Approaches to the Study of Organizational Buying

Behavior." In T. Green, ed., *Increasing Marketing Productivity*. Proceedings of the 1973 AMA Conference, 203-206.

——. "Industrial Market Segmentation Under Conditions of Intra Organizational Heterogeneity," Wharton School Working Paper, December 1976.

——. "Organizational Buying Centers: A Research Agenda." In Gerald Zaltman and Thomas V. Bonoma, eds., *Organizational Buying Behavior*. Chicago: AMA, 1977.

Wind, Y., and R. Cardozo. "Industrial Marketing Segmentation." *Industrial Marketing Management* ⅃ (March 1974):153-165.

Wind, Y.; J. F. Grashof; and J. D. Goldhar. "Market Based Guidelines for the Design of Industrial Products." Wharton School Working Paper, October 1975.

Wind, Y.; Stuart Jolly; and Arthur O'Connor. "Concept Testing as Input to Strategic Marketing Simulations." In E. Mazzie, ed., *Proceedings*. 1975 AMA Conference, 120-124.

Wind, Y., and E. Lotshaw. "The Industrial Customer." In S. H. Britt, ed., *Marketing Handbook*. The Dartnell Corporation, 1973.

Wind, Y., and John G. Myers. "A Note on the Selection of Attributes for Conjoint Analysis." Wharton School Working Paper, January 1977.

Part 3

The STI Product

Introduction

In the first paper in this section, Edgar A. Pessemier builds on ideas from the previous section to address the issues of STI "product" planning. Thinking of STI as a product that must be designed, developed, packaged, and promoted on the basis of identified user needs is an innovative concept. Often, STI is treated as an entity that is "naturally" produced through research. This leads to a production-oriented focus that treats information only as something to be "distributed." The same thinking, as applied to consumer and industrial products, has often led to marketing failure. There is, therefore, reason to believe that a purposeful product planning concept may have far-reaching implications to the marketing of scientific and technical information.

Consistent with trends in other fields, there has been an increased interest in marketing with the utilization of research. Marketing spokesmen have been addressing issues concerning the effective and ineffective use of information (e.g., Barabba, 1978; Greyser, 1977). While the record of information or research utilization in marketing leaves something to be desired, the marketing profession is considered by people in other professional contexts to be relatively advanced in its understanding of information utilization processes and in its information utilization practices (Echols, 1973).

The second paper, by Gerald Zaltman and Rohit Deshpande, identifies, in the form of guidelines, various practices that are generally present when there is effective use made of marketing information. A central theme among the various guidelines is

that the producer and user of research—the information product
—interact and understand each other's needs, abilities, and constraints. The development of information products requires a
user orientation on the part of the information producer *and* a
producer orientation on the part of the information user. While
the authors focus on information developed, communicated,
and used to solve a specific marketing problem, the basic
philosophy applies to less mission-oriented information
products. It is suggested that the design of both information
development and transfer systems or services should incorporate
many of these guidelines.

References

Barabba, Vincent. "Knowledge Utilization—Differences Between Public/
    Private Agencies." Paper presented at the Research Utilization Con-
    ference, University of Pittsburgh, September 20-22, 1978.

Echols, James R. "Introduction to the International Conference on
    Making Population/Family Planning Research Useful: A Communica-
    tor's Contribution." East/West Communication Institute, East/West
    Center, Honolulu, December 3-7, 1973.

Greyser, Stephen A. *Improving Business and Academic Collaboration on
    Research in Marketing.* MSI Special Report No. 77-100. Cambridge,
    MA: Marketing Science Institute, 1977.

W.R.K.
G.Z.

6
Product Planning for Effective Scientific and Technical Information

E. A. Pessemier

Perspectives and Definitions

In the usual market context, a product (service) is the object of economic choice and consumption. The use-value of such objects depends on three factors: the characteristics of the product (and associated services), the characteristics of the user, and the conditions under which it must be used. Since it is difficult to influence users or conditions of use, attention is commonly focused on improving the assortment of available products. This objective can be achieved by expanding the set of useful products and by improving the characteristics of the available products. In this sense, *discovery* and *design* become the principal areas of analysis and action.

When scientific and technical information (STI) is being considered in a product context, it is tempting to minimize the relevance of purposeful discovery and design. Many scientists and technicians believe discovery is serendipitous and the nature of each discovery defines the essential characteristics of the related STI product. If this view is correct, the common analytical approaches to product management are useless. The remainder of this paper will discuss the dysfunctional results of accepting this passive approach and the promise of adopting a product management approach to STI.

Supply and Needs, Availability and Use

A management approach necessarily implies the deployment of resources to efficiently attain one or more useful goals.

Scientific and technical information (reported findings) adds to the stock of useful knowledge and encourages further additions to the stock. In a sense, the scientific and technical enterprise moves toward a retreating and ever-widening horizon. Each new addition to the supply of knowledge produces a larger need for knowledge. Unlike most objects of choice and use, each unit of supply is unique and can benefit many users in varied ways over very long periods.

Scientific and technical information is consumed by obsolescence and neglect, not by use. Furthermore, the use of scientific and technical information can be viewed in several ways. In the most trivial sense, use can be equated to publication and/ or exposure. At a deeper level, information is used when knowledge is put to work for some constructive purpose. Finally, use occurs when information is employed to create new knowledge. Focusing on the latter two aspects of use, it is easy to conceptualize the value of any new information product, as shown in Figure 6.1.

This schematic clarifies the often-neglected fact that scientific and technical information must ultimately be judged by its contributions to the future stream of constructive applications. In this regard it is useful to note that some applications will have greater social value than others. If new knowledge is not solely serendipitous, then the scarce resources of the scientific and technical community can benefit from information about the nature of social needs and the technical possibilities for satisfying these needs.

Directional Decisions for STI Products

The foregoing analysis describes some important characteristics of scientific and technical information. Some similarities and differences of consumer and industrial products and STI products have been briefly mentioned. It is also useful to recall the nature of new product development as it is carried forward in the business community. There, the first task is the identification of user needs for novel or improved products. This scanning of the market and related socioeconomic environment raises the sensitivity of product development personnel to im-

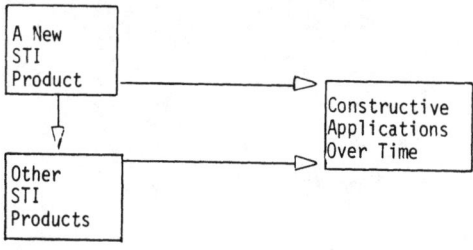

FIGURE 6.1: VALUE OF A NEW INFORMATION PRODUCT

portant unfilled needs. If few technical problems are expected, scanning for opportunities, formulating imaginative designs, and testing alternative product solutions are the principal new product activities. When product needs are well understood but suitable technical approaches are not readily available, the feasibility of developing a product, the cost of finding a satisfactory solution, and the value of success become the dominant concerns. In these latter cases, the resources of an industrial laboratory must be rationally deployed with proper allowance being made for the expected returns from their work. Decisions in this realm call for information about both the state of technology and the state of the market place.

Scientific and technical information fits the above conceptual framework. Without appropriate information about such things as the prospects for success, the effort required, and the expected value of results, the efforts of scientists and technologists will become aimless and ineffective. Although fortuitous discovery will continue providing a very important share of all new knowledge, there is ample evidence that knowledge responds to need and that available resources are sensitive to the nature of the needs (2). It is also true that the scientific and technical community responds to perceptions about emerging opportunities for useful new knowledge and to the interest of other scientists in the potential results of their work. Some of the essential relationships can be represented schematically, as in Figure 6.2.

The accumulation and dissemination of information that can *guide* scientific and technical work is one of the weakest links in

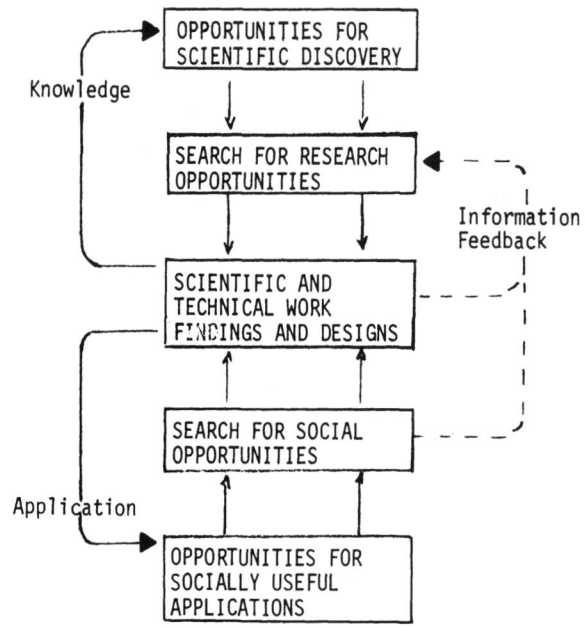

FIGURE 6.2:   INFORMATION AND THE SCIENTIFIC PROCESS

the effectiveness of the system outlined in Figure 6.2. It is not easy to appraise either the feasibility or social value of achieving a given result. The area of "big science" provides a number of conspicuous illustrations. The engineering and economic aspects of large-scale projects such as the development of fusion power raise serious scientific and political issues. Here, the cost of developing and testing two or three alternative technical solutions can run two to four billion dollars a year for several decades (1). In this type of effort it is not uncommon to purposefully move ahead with construction and testing efforts before design feasibility or efficiency has been demonstrated. This strategy should rapidly locate significant problems, suggest potential solutions, and add to the body of related knowledge. On the whole, however, this brute force approach involves high costs and risks.

At the other extreme, "little science" is a cottage industry that principally serves personal and institutional needs. The investigator has little if any directional guidance aside from the values of journal editors and the reviewers for agencies who may provide modest financial support. The quality of research results is highly variable. In addition, individual investigators seldom have the staying power necessary to make important contributions to some areas of science and technology.

The most serious question facing those interested in improving the STI product concerns how to identify priority needs, how to evaluate the scientific and technological promise of these areas and, finally, how to encourage qualified individuals and groups to take up the work. In my own field of interest, I find little encouragement. Huge areas of practical significance to economic and social welfare receive little or no sustained effort by qualified investigators. These large opportunity losses seem to be principally the product of ignorance about the nature of needs, of a lack of faith in the feasibility of finding solutions and, perhaps most important, of an unwillingness to look at potentially controversial areas of institutional, community, and individual behavior.

STI Product Management Compared to Industrial Product/Service Management

The time phases of industrial product development provide a useful frame of reference for comparing the STI product environment to that found in industry. Within each of these phases, important similarities and differences can be readily observed.

Search and Preliminary Screening

Consider the individual investigator or decision maker contemplating a scientific or technical problem. It is likely that the problem is receiving attention because it was uncovered by routine scanning of social and technical needs, or by a recognition of an opportunity to extend an established body of knowledge. If the need or knowledge is judged to be relevant to the individual or his institution, exploratory research or preliminary analysis will be in order. In some cases, it may involve further assessment of the need. In other instances, further study of the chances of success may be called for.

At the conclusion of these activities, the expected rewards from work on the problem can be appraised and compared to the expected rewards from other opportunities. Also, the likely behavior of the person or institution who will provide the required resources for continued work must be considered. This complex of influences produces a queue from which potential new research (and development) projects will be drawn. The explicit content of this prioritized list will be strongly conditioned by perceptions about each of the elements described above (needs, probabilities of success, potential support) and the personal values the investigator or decision maker places on the possible outcomes. In an indirect way the list is conditioned by the competence of the worker(s) and the current work load.

This example underscores the importance of directional information. Within a fixed set of resources, better data about needs and possible ways to satisfy these needs will influence the content and structure of the project queue. Furthermore, directional information can shift the capabilities and capacities of the work group, thereby changing the queue content, its structure, and its rate of service. Just as the success of an industrial laboratory is shaped by staffing and project search decisions, the type and rate of science and technology findings will be heavily influenced by the capacity of the research community and the tasks that it undertakes. Directional information is needed to guide the recruitment and training of scientists and engineers and to help them direct their efforts.

Development

Once an investigation begins, the substance of the output emerges at a pace that is controlled by the nature of the problem and the weight of resources applied to achieve the desired results. (As in the case of industrial product development, the needs of the "consumer" shape the nature and form of the findings.) If an optimal solution is the objective, not simply a workable solution, the time required may sharply increase. The same effect may be observed when a general solution is sought, not just an answer to the specific problem at hand.

In a similar manner, if results are directed toward engineers, more detail, illustrative material, explicit test results, and the like may be appropriate. Audiences with a more theoretical bent may require carefully presented logical foundations and mathematical development. Some projects will call for elaborate equipment, extensive data collection, or other elements that are costly in terms of time and money. Other projects call for little more than modest reference and computational aids. When the nature of the research task is coupled to the needs of research consumers, the development task becomes meaningful. Substance and form become two inseparable parts of the development effort. In general, these two aspects are more easily recognized in industrial product development activities than in STI product development activities.

Testing and Introduction

The testing that goes into industrial product development prior to market introduction is designed to prove the worth of a product prior to its wider market exposure. The risks to both seller and buyer are significant factors. Furthermore, it is typically important to appraise both the initial acceptance and the longer-run value of a product. How many potential users can be expected to try the product? Having tried the product, will they continue to find satisfaction from its use over a long period of time?

The judgments of colleagues and journal editors frequently constitute the primary testing function for scientific and technical knowledge. If the prospective contribution is rejected at this level, few will know about the work. On the other hand, if findings appear in the formal literature, the contributor receives a wide audience and the "endorsement" of an authoritative sponsor. Although the performance of this primary testing method is far from uniformly satisfactory, its importance cannot be easily minimized. Furthermore, the demands it makes on the form and substance of the research findings it disseminates cannot be easily adjusted to the specific needs of contributors or users of new knowledge.

Once an STI product has been introduced, the reputation of the researcher and publication will influence initial "adoption" or first exposure. Beyond this point, however, the general value of the contribution becomes progressively more important. If the new knowledge is significant and widely applicable, it will be cited in the work of other researchers and diffused throughout the scientific and technical community. If materials do not receive wide distribution when first published or are noted principally by individuals who do not contribute to the literature, the diffusion process will be hampered. Therefore, the right type of early exposure has a strong influence on the rate of adoption and diffusion.

Like the primary and secondary markets for commercial products and the associated distribution channels, STI products have *primary* audiences and media and *secondary* audiences and media. To efficiently provide various types of potential user markets, consideration must be given to both the form in which the knowledge is delivered (product design) and way it is delivered (channel strategy). Part of the testing and introduction process in industry is devoted to exploring alternative product designs and marketing mixes for users with different needs, motives, and resources.

On the whole, not much effort or imagination has been devoted to alternative designs and delivery strategies for new knowledge. Typically, the contributing researcher does not have the same dependence on user satisfaction found in commercial product development. This independence of the producer of knowledge has positive aspects but it also has substantial drawbacks. Furthermore, editors seldom play the active user advocacy role commonly found in consumer markets. Funding agencies can have a strong directional influence, but their impact on the form and distribution of STI may not be as significant.

The STI Product Line Assortment Problem

Thus far, our attention has focused on single, one-at-a-time contributions to knowledge and how this knowledge is delivered. A somewhat more complex topic concerns related collections of knowledge. Are these collections sufficiently focused and complete to assist users? Can a collection's value be enhanced by influencing the form of presentation and the range of topics covered, or must gaps be filled by encouraging new work? A variety of serious questions arises in these regards. First, should gaps be recognized as part of the directional effort needed to guide new work? Second, can important problems and user-specific needs be recognized that can guide the preparation of suitable collections? Finally, what institutions and individuals can assume useful roles in encouraging productive new work and the effective packaging of existing knowledge?

Clearly, book publishers and journal editors are decision makers who can make a positive contribution. Their efforts appear to be largely opportunistic and fall far short of the ideal. Handbooks, special collections, and annual reviews, particularly those found in the social and administrative sciences, have little lasting value. One is tempted to conclude that most academic and commercial incentives encourage low quality output that may crowd out higher quality, well-focused materials. The question of incentives and quality control need more attention. Additional comments about this appear in the concluding section of this paper.

Some Patronage Issues

In a commercial context, patronage concerns the degree to which a buyer or user satisfies recurring needs by selecting the same product, producer, or distributor. In the STI context, this vital concept is not very well defined. Does a scientist patronize specific libraries, publishers, abstracting services, intellectual leaders, and gate keepers, or collections of journals in a way that is analogous to how he may return to a brand or store that has a history of satisfying his consumption needs? In a very broad sense, the answer is yes, but unlike producers and distributors of products, less is understood about STI needs and the extent to which they can be satisfied. The burden is on the user to find what is useful among largely unstructured collections of knowledge—knowledge that typically is not generated, packaged, or presented with the user in mind. In STI, most users face a severe case of "scrambled merchandising." Furthermore, the STI "retailer" may not be greatly concerned about maintaining a large, active group of patrons. Instead, these providers of information tend to have a stronger need to maintain good supplier relationships, in part because many patrons are potential contributors.

In contrast to most commercial contexts, little is known about the patronage of single or multiple product outlets or which STI products are noted and/or used by various types of individuals and groups. In the case of some professional society proceedings, the rate of exposure to potential users is doubtless very low. For some journals, say *Psychometrica*, the rate of exposure per article may be relatively high within the journal's small and active audience.

In addition, not much is known about why and how STI products are used. What fraction of the time is the potential user scanning to add to his fund of generally valuable knowledge, to add to his professional expertise, and/or to help solve a specific problem? What part of the exposure to STI is to retrieve old knowledge and what part to locate new knowledge? These issues bear directly on the design and delivery problems for STI. More serious research on these topics could improve the efficiency of information systems.

Is Anyone in Charge? Is Anyone Listening?

The foregoing description of the STI product problem tempts one to conclude that the process is not efficient and is not very responsive to *user* needs. It may, however, be more efficient and responsive to *contributor* needs. The needs that it best serves among contributors, however, appear to be principally personal and organizational. As long as STI output represents an important input to the prestige structure and personnel management process of professions and organizations, contributors are unlikely to be strongly influenced by user needs. Furthermore, this supplier bias encourages the publication of large amounts of STI that are largely worthless to the masses of STI users.

A second and related issue concerns the role of funding agencies with short term explicit needs. Here, investigators' efforts are channeled into work which is unlikely to produce findings of value to wide classes of users. In part this may be due to the proprietary nature of some investigators but more often it is the result of a solution-finding orientation as compared to a knowledge-finding orientation. Expanding the availability and generality of institutional research programs could add an important degree of scientific and technological utility of much current work.

References

1. Rose, David J. and Michael Feirlag. "The Prospects for Tuscon." *Technology Review* (Dec. 1976):20-43.

2. Utterback, James M. "Innovation in Industry and the Diffusion of Technology." *Science* 183:620-626.

7

Increasing the Utilization of Scientific and Technical Information

Gerald Zaltman
Rohit Deshpande

The utilization of scientific and technical information (STI) is an issue that is at the very core of the marketing of any form of knowledge. Information that has been produced and is either not utilized or is underutilized largely defeats the purpose of knowledge production—to create knowledge and an awareness of its availability to individuals so that they can put it to optimum use. In marketing, a field of inquiry that is supposedly better equipped to handle utilization issues than other less applied social science fields (due to the profit motivation for information collection and use), only about a third of research information that is produced is estimated to be used. Research directors in leading advertising and research organizations indicate that it is unclear how much of that third of information is used *effectively*.

The problem of information utilization is no less acute in other areas. In the last twenty years the knowledge industry has grown dramatically, with annual federal expenditures on research and development climbing to $30 billion in recent years. Federal programs supportive of applied research on specific social problems (education, energy, mental health, housing and urban development, crime) have themselves approached annual funding levels of $20 million. Until very recently, however, the public commitment to knowledge production has not been matched by a serious concern with its practical uses. In fact, the utilization of scientific and technical information may be the Achilles' heel of science. One of the leaders in the developing field of knowledge transfer, Dr. Howard R. Davis, recently offered the following comments (Davis, 1976, p. vii):

> The knowledge industry must surely be considered one of the majors in the nation. In research and development alone, the Federal investment amounts to some $20 billion annually. Curiously, it is an industry where astonishingly little attention has been dedicated to the marketing of its products. . . . In the past 20 years the number of citations in the knowledge utilization field has grown from some 400 to an estimated 20,000 plus. If you venture into this literature in a quest for guidance, you stagger out reeling. The field abounds with assertions, conceptual models for analysis, and contradictory observations.

This paper is intended to provide several guidelines that can enhance the utilization of scientific and technical information. These guidelines should be of concern to both producers (researchers) and users; both groups have responsibilities for enhancing the usability and usefulness of research and both groups have a great deal at stake in improving the information utilization process. The guidelines to be discussed and derived here are supported by several sets of observations. These include comments made to the authors by research directors of major marketing research agencies, remarks made by major users of research in consumer and industrial industries, findings from studies outside of marketing that have addressed the issue of information utilization, and the observations of the authors in their consulting and research activities.

Information Utilization Defined

Information utilization is defined here as the process by which users' needs are determined and communicated to producers (researchers), leading to information designed to meet these needs, and eventually to new knowledge based on information that is passed on to users who apply it to answer their needs. Figure 7.1 is a schematic representation of this definition. Each basic element in this definition represents an explicit area that should be of major concern to producers and users of STI. In the discussion to follow, it does not matter whether the "producer" is an in-house research group or an

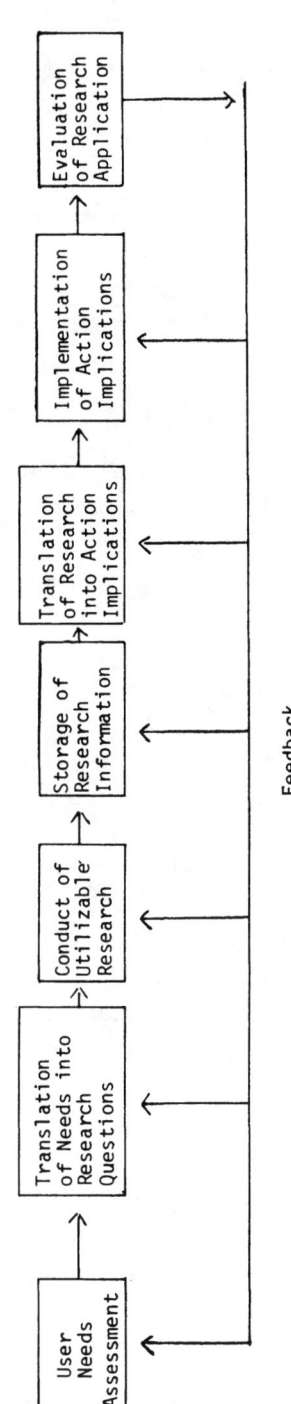

FIGURE 7.1: THE MAIN ELEMENTS OF THE STI UTILIZATION PROCESS

information supplier contracted by the user. The guidelines apply to both situations. These guidelines are organized according to the seven major components of the information utilization process.

Although most of the illustrations described in this paper come from marketing firms, the management of knowledge in general, and of scientific and technical information in particular, transcends the boundaries of corporate entities and applies to society and its research-production and research-use functions as a whole.

User Need Assessment

User need assessment is the process of determining what images or ideas the ultimate users of research, such as production supervisors, consulting engineers, new product ventures groups, and so forth, have in mind about what *should* be. What information do they believe they should have but do not now have to make decisions? A proper understanding of the desired levels of information that users want is absolutely essential if research producers are to be effective in satisfying user STI needs.

There are several guidelines that concern user need assessment.

Guideline 1. *A mechanism should be established to facilitate the continuous assessment of user needs.* These needs may or may not be clearly articulated by users. For example, an experiment being tried by a firm in the apparel industry involves having personnel from their marketing research department meet bimonthly with various management personnel to discuss general research needs exclusive of any specific ongoing research project. Interaction with the research pesonnel appears to help managers refine the expression of their needs. This interaction also appears to be giving research producers a better sense of their user colleagues' general information needs. Additionally, continuing interaction between producers and users is likely to facilitate the precise pinpointing of research needs with a growing comprehension of the user's culture by the information producer and vice versa.

Guideline 2. *The user need assessment activity should include attempts to forecast information needs which do not exist now but are likely to develop in the near future.* The forecasting of information needs is essential if the time gap between the actual experiencing of a need and the availability of usable research to address that need is to be shortened. All too often marketing decisions, for example, must be made well prior to the earliest possible time at which relevant research could be gathered. It is difficult, of course, to know which of the predicted research needs will not materialize. The cost of commissioning research for anticipated needs that do not ultimately materialize must be compared with the expected cost to the firm of not being able to respond effectively to a need for the period of time during which research is being conducted.

Either users or researchers or both in cooperation should be involved in the forecasting of research needs. This leads to Guideline 3. *Need assessment technology must be further developed and taught.* Researchers and users should become familiar with the various techniques to use with various market indicators that can identify both current and future research needs. Need assessment technology is not as well developed in the STI area as it is in other fields such as education.

Guideline 4. *A need assessment program should identify user-initiated efforts to meet expressed needs.* In the scientific instruments industry, manufacturers appear to rely very heavily upon users of scientific instruments to formulate their own solutions to problems. In this instance manufacturers appear to be very tardy in becoming aware of and assessing user needs (von Hippel, 1978). An early identification of user needs would help producers of scientific instruments respond more successfully to their needs in a marketing sense and users would benefit from producer expertise much earlier.

Being attentive to user needs is important where information users and producers are within the same organization. It is not uncommon to find managers bypassing available research staff in the same firm or on retainer because managers believe they can obtain adequate information themselves. In-house research staff in many industrial firms keep a constant watch for this. First, perhaps, because it lessens the research producer's importance to the firm if he is not consulted or used, and second, because managers often cannot obtain adequate data even if they believe they can (or at least they cannot obtain it as efficiently as the research staff). Yet even if users do obtain the required information, it is inappropriate that they spend their time in duplicating activities for which professionals have been hired.

The Translation of Needs into Research Questions

Once a need has been identified among a user group, it is necessary to state this need in terms of a researchable question. This involves a practice found among many marketing researchers who cast marketing problems or needs into theoretical issues. For example, in the field of marketing, a problem about repetition in advertising might be expressed as an issue in learning theory. Existing knowledge about repetition in learning might provide clues as to whether advertising should be repeated at regular or irregular intervals. The application of learning theory in other contexts might provide still further insights into the marketing problem. Similarly nonuse or underutilization of journal articles by chemists, for example, might be construed as an issue of information overload. Viewed in this way new journal formats may be designed (Terrant and Garson, 1977).

Translation of a need into a researchable question and identifying relevant disciplines or subdisciplines is essential (1) to bring all plausible sources of information to bear on a need, (2) to investigate the need properly, and (3) to elicit the attention and interest of researchers and persons who are not oriented toward applied research but who are important contributors to the thinking of an area. This brings us to a number of guidelines.

Guideline 5. *Researchers and users should have a shared responsibility for translating problems or needs into research questions.* Users' needs must be converted into research questions so they can serve as a useful guide toward utilizable research. Ideally, researchers and users should interact on a face-to-face basis in formulating research questions. That is, persons designing new information systems should do so collaboratively with the users of those systems. This indicates a need assessment-research formulation feedback mechanism as shown below.

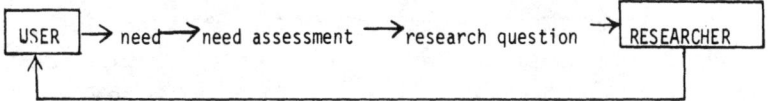

Guideline 6. *Every user need should be explicitly related in a chart or table to one or more research questions.* Although this is a standard procedure in many organizations, it is often not done in others. The chief value of this guideline is three-fold: (1) it ensures that a need is not overlooked; (2) it ensures that a control mechanism exists that can later be referred to by persons not intimately connected with the specific research to determine what needed to be researched and how this was formulated for the purpose of the research itself; and (3) it ensures that there is common agreement (or at least provides a basis for determining there is not agreement) on which research questions address which user needs. This last point may need clarification. The authors have witnessed numerous instances in which researchers do not agree on which of two or more questions proposed in a questionnaire answers a specific user need. The kind of answer provided in response to a user need might vary substantially depending on which of the questions at issue is selected. For example, if an industrial firm is interested in learning what kind of person is likely to *actively* stimulate word-of-mouth communication and what kind of person is likely to be *sought out* as a source of technical product information, they would ask questions about opinion leadership. However, different sets of questions would be asked to identify two (possibly) different kinds of people. One of the authors recently encountered a situation in the chemical industry in which the user firm had used the set of questions about stimulating word-of-mouth communication (mentioned above) to define the more passive people sought out by others for product information. Had the research firm clearly identified which questions were specifically intended for each of the two kinds of people the user firm was concerned with, this mistake could have been avoided.

Guideline 7. *Users and especially researchers should be trained to translate information needs into researchable questions.* Great differences exist among people in their abilities to perform this function. A great deal of skill is needed for the seemingly simple process of translating users' needs into researchable questions. Both researchers and users should be involved in training sessions to learn how to perform this function better. Such training is feasible and has been done elsewhere (Argyris and Schon, 1974). These techniques involve having users engage in exercises involving the translation of general needs questions into specific researchable questions. For example, in the field of marketing, the issue over the role of a purchasing agent in a buying committee concerned with a major equipment purchase is translated into specific questions about the mechanisms utilized to disseminate information about a particular product or service that the company is considering purchasing.

Guideline 8. *User needs should be assessed for their researchability.* Not every need is readily translated into an easily researched question. Again in the marketing context, learning who the gatekeepers are for scientific and technical information is important for the development of a campaign to stimulate word-of-mouth communication relying on key purchasing agents, consultants, and other technical specialists. For a variety of reasons this is a very difficult question to ask in a way that ensures validity and reliability. It is important that users understand at this stage which of their needs are difficult to research. This minimizes later misunderstanding and conflict.

Conduct of Utilizable Research

There are many ways of ensuring the usefulness of STI by following certain guidelines during the actual research process. A simple conceptualization of the research process is contained in Figure 7.2. Several guidelines are suggested below.

Guideline 9. *Users should inform researchers of the kinds of decisions that will be made on the basis of the research.* This may help researchers perform their work. There may be a reluctance for users to convey this information when the data collection is being performed by an outside firm. This reluctance may be based on the assumption that the fewer the people who know about a contemplated major change in corporate policy or strategy, the less likely it is that competitors will become

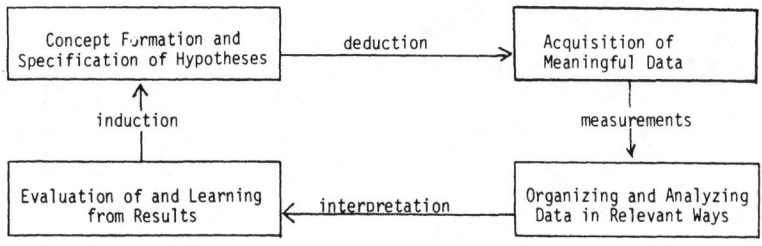

FIGURE 7.2: A SIMPLIFIED PARADIGM OF RESEARCH PROCESSES

aware of it prematurely. Whether it is for this reason or for other reasons, there may be a loss of information value if the kinds of decisions to be made are not communicated to researchers. When researchers do have this information, they can formulate basic questions, the research design, and the research instruments in more relevant ways. For example, if the researcher knew that the data being gathered about which procurement managers are loyal brand purchasers would eventually be used for a decision to either launch or drop a product, the information being sought could be made more specific to this potential decision. Additionally, the research agency may be able to bring in data from earlier research that would add to the findings of the study being currently done.

Guideline 10. *Prior to the actual collection of data, the research team should meet with users and review the research instruments.* This goes beyond consultation about research questions and requires a review of how these questions are to be expressed in a mail questionnaire, interview schedule, or other data-collection instrument. Probably the aspect of the research process that is easiest to criticize in survey research in the social sciences is the questionnaire. Questionnaires, no matter how skillfully designed, are extremely vulnerable documents if they go beyond very simple "nose count" issues. It is especially important, then, to maintain a constructive atmosphere in group meetings when reviewing research instruments. Researchers must also be prepared to discuss with users their rationale for selecting specific methodologies, e.g., why a mail questionnaire rather than personal interviews, why a cluster sample, and so forth. An illustration of this in high energy physics may be found in Libbey and Zaltman (1967).

Guideline 11. *The researcher should also consult with users about the form in which they want the data presented to them.* Formats desired by users may differ. Depending upon the time and purpose of information gathering, some scientists may prefer short summary articles in contrast to lengthy articles. The investigator should consult with users about how the data can be most meaningfully presented. The format in which data are reported is especially important when making a presentation to a group of users who may vary individually in how they respond to tables, charts, and figures. (Zmud, 1978)

Some users are easily "turned off" by complicated tables or tables that display considerable information, while other users become quite involved and stimulated. It is useful to create dummy tables, charts, and figures prior to the completion of the research and ask a sample of users if they find the particular formats appropriate. It may also be relevant to present data in a standardized format into which further STI research might be introduced. This will also facilitate the use of this information by other persons who were not present during the process of the initial research activity.

Guideline 12. *Researchers and users should consult with each other during the interpretation phase of the research regardless of who has the responsibility for the interpretation function.* An extension of this guideline suggests that *a clear designation should be made at the outset of the research about responsibility for the interpretation function.* Most readers probably know of unfortunate situations in which researchers have assumed users will interpret the data while the users have expected the researchers to provide the interpretation. The authors are familiar with several situations of this nature involving contracted research that have only been resolved by bringing a third party to provide the interpretation. Regardless of how responsibility for this function is assigned, it is very desirable that researchers and users discuss jointly all interpretations that are developed. Researchers can provide guidance about potential restrictions that the research methodology imposes on the interpretation of data. Users can provide additional insights about the actual contexts the interpretations relate to which may affect the interpretation.

Some of the most successful research projects the authors have observed in various social science research studies have involved simulated interpretations of data *prior* to the actual collection of information. Fictitious data are developed and presented in the anticipated format of the final tables, charts, and figures. Researchers and users then brainstorm the meaning of the data. This approach invariably results in modifications of the research instruments and, in some instances, alterations in the research design. This exercise seems to enhance immensely the usability of the final data.

Guideline 13. *It is desirable to have an information utilization expert assigned to research teams.* Many federal funding agencies are beginning to require this in projects they support. When this responsibility is formally recognized and assigned, there is a much increased likelihood of guidelines, such as those discussed in this article, being followed.

There has been considerable success in the very few instances the authors are familiar with in which a manager of a research team is formally assigned the role of information utilization expert. In all instances, this person has been the director of research in the firm. A modest-sized marketing research firm has recently hired a utilization expert to serve this function simultaneously in all their numerous research projects. Although it is too soon to draw any conclusions about the success of this experiment, the firm reports its early experiences to be highly satisfactory.

Guideline 14. *It is important to examine the research plan for ethical considerations before implementing it.* This is an issue that is frequently overlooked in marketing research activity and perhaps in the STI contexts of concern to readers of this volume. The result is that there are several complaints by research respondents that the questions they were asked were stressful to answer or that the research involved manipulation or deception. Both users and researchers need to be aware of and guard against the ethically questionable issues a research approach may raise. This guideline is important not only for the protection of respondents, users, and researchers, but also because the way in which particular ethical issues in research are addressed may seriously affect the quality of the data obtained (Tybout and Zaltman, 1974).

Storage of Research Information

Research results should be catalogued and stored in an easy-to-use information system. Ideally, this would involve research conducted by other agencies being reported in the public literature as well as research commissioned by the user firm. There are many government and private information retrieval systems that do this on a contract basis for specific topics. However, many firms find it helpful to maintain their own specialized system as well.

Guideline 15. *User needs should be one of the classifying criteria in a firm's research information storage system.* It is important for both researchers and users to be able to retrieve past information on the basis of the user needs and research questions investigated. Most information systems use completed reports and specific findings as the main or sole unit of memory. This is a source or message orientation. Often the potential users will not know exactly what sources or messages they need. This results in a broad search which may create information overload. Relying on user needs and research questions creates a more user-oriented information system. Such user-oriented information systems appear to be used by marketers with much greater satisfaction than the source or message-oriented system. Since the purpose of research is to generate information that will be useful to users, it seems logical that the retrieval system cataloguing should be arranged to facilitate user access.

One way of classifying information with user needs as classifying criteria is to organize documents and data on the basis of the questions the research answers. Consider an example drawn from marketing. Instead of filing a report, article, or other information about changes in the battery market under the label "battery market," it should be filed under "What are projected auto battery sales between 1978 and 1983?" and other questions the document answers. Management needs exist in the form of fairly specific questions rather than in terms of key words or document titles. Information systems should reflect this.

Translation of Research into Action Implications

Perhaps the single most difficult question to answer in the area of research utilization is "So what?" What does a research finding mean for a practitioner or decision maker? The need for such translation abilities is great. Social science research textbooks generally do good jobs in teaching the technical aspects of conducting research. At the same time, only rare instances can be found in which efforts are taking place to teach the process of translating research results into practice (Zaltman and Burger, 1975). If any stage in the research utilization process is more difficult or troublesome than another, it is this stage. Many research users complain that their research departments or contractors either refuse outright to translate the research they gather or that they only participate in the translation process reluctantly. It is understandable that users want assistance in this process and that researchers are often very hesitant to engage in it. Translating research into action is a very risky enterprise. It is easy to be wrong. Most research can be subject to multiple interpretations and a plausible alternative to an interpretation and translation that failed can usually be generated. Monday morning quarterbacks may have a field day. Yet the whole purpose of the research endeavor is that it produce recommendations for action. To do otherwise would be to simply engage in theoretical or academic considerations.

Guideline 16. *One or more users in an organization should be trained in translating research results into action implications.* Persons skilled in this task often serve as communication links between other users and researchers. Persons skilled in translation can generate many more plausible alternative implications of data than persons without these skills. This increases the likelihood, for example, of a manager selecting the most appropriate implication to follow in developing policy. User groups who have such a person appear to make far better use of far more research information. This person should not displace translation responsibility among researchers and users but merely facilitate it.

One of the authors has been involved in sessions designed to train product managers to be more innovative in interpreting research data. This has involved using models of buyer behavior, decision trees, role playing, and other tools and techniques for generating action ideas from marketing research. It is evident that asking what data mean in light of different models, different schools of thought, and so forth does help managers generate more ideas. However, it is also evident that the task of translating research results into action is not always easy. Moreover, the product managers just referred to displayed very different responses to the training efforts. Some seemed to improve their skills quite substantially. Others had only marginal improvements.

Guideline 17. *Where possible, action implications should be linked to corresponding research questions and the original user need from which it stemmed.* This guideline ensures the consideration of all user needs identified at the outset of the research. If there are research questions and user needs that do not have action implications associated with them, attention should be directed toward addressing those needs. Of course, there will usually be action implications not directly associated with predefined consumer needs. Good research will often produce serendipitous findings.

Implementation of Action Implications

The general issue of implementing research implications is very understudied. It is one thing to know what a given research finding suggests in terms of a new design for a scientific instrument. It is quite another matter to be able to go ahead and do it effectively given resource contraints among producers or users.

Guideline 18. *Researchers should be consulted by users in the implementation process.* Many new ideas work only if the developer is present to oversee their implementation. The presence of a researcher is desirable where adaptation of his basic ideas or findings is necessary to fit a particular user requirement.

An example from marketing further illustrates the desirability of having the producers of information available. An advertising agency that one of the authors has worked with regularly reviews their implementation plans with their research staff. In a recent instance, a user-generated action implication of a research finding called for the development of a luxury line of prefabricated vacation cottages designed to appeal to a very high income category. Among the alternative ways of implementing this action was the choice of acquiring an existing line of luxury prefab designs from a firm that happened to be dropping such a line, or to design their own. When presented with this information, one of the staff researchers was able to add information from secondary data sources that had been obtained but not placed in the research report because the data were originally considered irrelevant. This particular item of information made it very clear which way of implementing the action implication was most likely to succeed.

Guideline 19. *Researchers should be sensitive to barriers that users may face in implementing actions suggested by research.* It is not always possible to do what might appear to be an obvious way of putting an action decision into effect. For example, financial and nonfinancial resources may not be available to a user. Personnel may simply lack the skills or experience to undertake a particular new activity and funds may not be available to train the salesforce or acquire personnel with the requisite skills. Also, there are certain political and idiosyncratic factors operating among key managers that dictate the choice of, say, a way of implementing a new product decision even though research evidence does not favor that approach. Understanding these and other barriers may prevent researchers from getting frustrated because they think their work has been directed at seemingly "foolish" users.

Evaluation of Research Applications

The evaluation of research applications is as difficult as it is important. What time interval between the availability of relevant research and its first use is considered acceptable? How much elapsed time is acceptable between first use and widespread use? How is the intensity or degree of use by any user measured? What are the appropriate measures of benefit? Whose benefits should be measured? This set of questions could be greatly expanded. Fortunately, the area known as evaluation research has been making strides in the development of tech-

niques for evaluating various programs (Struening and Guttentag, 1975). There is much that we can learn from this area.

Guideline 20. *Every research project should have an information utilization evaluation component.* Responsibility for the
evaluation of the impact of the research may lie with the research team, with the users, or be contracted to a third party,
or some mixture of all three groups. Such an effort would yield
very substantial insight into the entire information utilization
process in a firm and should result in ideas for strengthening
future information utilization efforts. It is important that the
evaluation of the research process begin with the start of the
project rather than at the end of the project.

Guideline 21. *Information utilization evaluation should be
considered a part of the need assessment effort.* This brings us
back to the very start of the information utilization process.
The entire process can thus be seen to be a continuing cycle of
events that together constitute an effectively implemented
research-information utilization program.

Conclusion

Enhancing STI utilization is a very major concern in the user
community. It is fair to say that most research information is
not used or at least not used as well as it could be. The reasons
for this are manifold. For the most part, these reasons are not
related to carelessness or lack of interest among researchers and
users. Rather, the problems are rooted in differences in orientation between researchers and users, the complexities of the
research process, the realities of the users' environment, and a
myriad of other understandable circumstances. Several guidelines have been presented which attempt to address many of the
problems serving as barriers to more effective research use.
Undoubtedly the reader will find missing some guidelines he or
she has found important. Also, the reader will find some guidelines easier to implement than others. Overall, most guidelines
will be applicable to nearly all STI settings. The guidelines
represent the common experiences of many different persons
and organizations in marketing. It is hoped that readers may

benefit from these experiences by considering the application of the guidelines in this chapter to their own settings.

References

Argyris, Chris, and Donald A. Schon. *Theory in Practice, Increasing Professional Effectiveness.* San Francisco: Jossey-Bass Publishers, 1974.

Davis, Howard. As quoted in Edward Glaser, et al., *Putting Knowledge to Use: A Distillation of the Literature Regarding Knowledge Transfer and Change*, p. vii. Los Angeles: Human Interaction Research Institute and the Mental Health Services Development Branch, NIMH, 1976.

Libbey, M., and Gerald Zaltman. "The Role and Distribution of Written Informal Communication in Theoretical High Energy Physics." New York: American Institute of Physics, Report No. AID/SDD-1 (rev.), Report No. NYO-3732-1 (rev.), 1967.

Struening, Elmer L., and Marcia Guttentag. *Handbook of Evaluation Research.* Vols. 1 and 2. Beverly Hills, CA: Sage Publications, 1975.

Terrant, Seldon W., and Lorrin R. Garson. *Evaluation of a Dual Journal Concept.* Washington, DC: American Chemical Society, 1977.

Tybout, Alice M., and Gerald Zaltman. "Ethics in Marketing Research: Their Practical Relevance." *Journal of Marketing Research* 11 (November 1974):357-368.

von Hippel, Eric. "Successful Industrial Products from Customer Ideas." *Journal of Marketing* (January 1978):39-49.

Zaltman, Gerald, and Philip Burger. *Marketing Research: Fundamentals and Dynamics.* Hinsdale, IL: Dryden Press, 1975.

Zmud, Robert W. "Concepts, Theories, and Techniques: An Empirical Investigation of the Dimensionality of the Concept of Information." *Decision Sciences* 9 (1978):187-195.

Part 4

The Distribution of STI

Introduction

Marketing involves the distribution of products and services to consumers. Channel theory is that branch of marketing concerned with the design, operation, and evaluation of distribution systems. In the STI context, distribution systems consist of organizations and markets that function to transfer information from the producer to the consumer. In a very real sense, the "channel" is the meeting ground for producer and consumer.

In his paper "Channel Theory and Distribution" Michael Etgar addresses marketing channel concepts and their implications for the distribution of STI. In doing so, he discusses potential scenarios for the development of new marketing channels of STI. Alok K. Chakrabarti elaborates on the channel theme by considering channel selection strategies in the other chapter in this section. His paper discusses several implications of selected studies of the marketing of scientific and technical information. Together, the two papers provide several ideas that might result in improved distribution systems. As is characteristic of all earlier papers, Etgar and Chakrabarti express the need to consider the user in the development and implementation of channel strategies.

W.R.K.
G.Z.

8
Channel Theory
and STI Distribution

Michael Etgar

Channel theory is a specific branch of marketing thought that studies the development and management of economic institutions and agencies concerned with moving products from producers to users. Like other areas of marketing, channel theory is truly eclectic in its origin. It is based on contributions from economics (7), behavioral sciences (48) and operations research (2, 38), as well as inputs from comparative marketing (41), agricultural economics (18, 36), and communications theory (35).

The result is a broad analytical framework of models, hypotheses and research results. Channel theory attempts to address itself to several key issues:

1. How should channel performance be evaluated?
2. What is the economic rationale for the emergence of different channel structures and what is the relationship between different channel structures and performance indices (both in macro and micro terms)?
3. What are (if any) the general laws that dominate structural changes in distributive channels?
4. What are the implications of the tendency of channel institutions to operate as teams and develop as social systems?
5. What are the effects of existence of conflict situations among diverse channel members and how should conflicts be resolved and managed to ensure long-run channel survival?
6. How and what economic and behavioral tools could be used by channel managers to enhance channel cooperation?

At the micro level, interest in channels focuses primarily on management-related topics that deal with the potential responses different channel strategy mixes can generate in terms of market coverage and penetration, services performed, control potential, and costs. At the macro level, interest primarily focuses on the differential ability of channel systems to achieve socially desirable goals.

While the channel theory has been primarily developed from the analysis of the distribution needs of tangible products, its frame of reference has recently been applied also to distribution of services (10, 17, 28) and it may be applied to distribution of information as well. Information must be passed on to users as a tangible good. Time and space gaps between production and use have to be bridged and channel functions have to be performed. Economic decisions about the most efficient configurations of agencies that could perform these functions have to be made. As similar decisions are made regarding distribution of goods, conclusions of the channel theory can be applied to STI channel decisions.

The point of view adopted in this paper is *macromanagerial* (55). It embraces the position of a designer who is interested in advancing public welfare goals regarding the distribution of STI and the efficient performance of the STI system. The analysis is concentrated on channels for scientific information—those primarily directed to scientists and researchers. There are other channels that parallel and interact with the channels in science (51). For instance, Garvey et al. argue that the channels of distribution of STI to practitioners and applied users, although they may interact with science, have developed for the most part independently and are qualitatively different (21).

This paper will explore the potential contributions that channel theory can make to the improvement of distribution of scientific and technical information (STI) to the scientific community. Several basic tenets of the channel theory are reviewed and several implications for the distribution of STI are discussed.

Channel Design

Traditional economic and business analysis has tended to focus on the behavior of individual firms within the marketing channel. Channels of distribution have been perceived as loose coalitions of independent business firms, each bent upon maximizing its own set of goals with little regard for the goals of other channel members. Series of intermediary markets were assumed to be the primary vehicles of channel coordination between the different sets of channel members with price and the related tools of the market ensuring some degree of channel cooperation (15).

Recently, however, it has been recognized that often the intermediary market mechanism may be insufficient to ensure channel coordination. In many cases channel members make suboptimal market decisions (49, Ch. 1), the direct and indirect costs of the intermediary market mechanisms are too high, and such mechanisms may be also ineffective (15). Consequently, a total channel systems approach has become more accepted (5).

The advantages of a systems approach to marketing channels lie primarily in the area of planning, design, and control. If the channel can be treated as one system, goals can be well defined and proper channel designs can be selected. Channel performance can be evaluated and its operations planned and scheduled. These advantages of channel planning are pertinent to planning at both micro and macro levels.

Figure 8.1 shows the process involved in the design of a marketing channel. *Marketing service outputs* and other performance criteria are the starting point of the process and a basis for channel performance evaluation. *Allocation of* marketing functions among customers, producers, and channel intermediaries (or the commercial channel members), is the next stage of the process, and it results in a design of specific *channel tasks* to be achieved by the commercial channel members.

The channel tasks are achieved through the performance of marketing functions or flows in which various channel members participate. The channel organization that emerges or is planned in response to these demands is the *channel structure*. Channel designers will, therefore, strive to establish a channel organization that will perform best. Evaluation of channel performance is the end activity of the channel design process. It should be used as a feedback to affect decisions about channel tasks and structure.

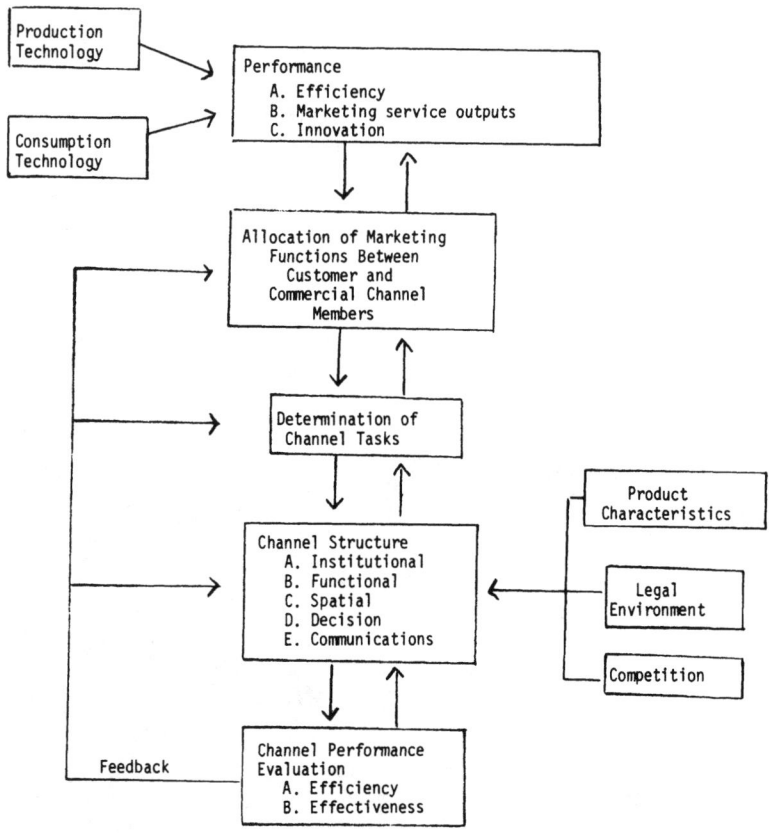

FIGURE 8.1:   THE PROCESS OF CHANNEL DESIGN

Channel Performance

The definition of a channel's performance is a complex and difficult task because of the absence of a unique performance measure. Instead, a variety of measures and indicators exist to reflect the multidimensionality of channel performance. Different groups within the channel may have different preferences and try to pull the channel in their direction. Consumers themselves may differ as to the utilities attached to the various services that the channel provides.

Channel literature differentiates between three broad types of channel performance: those pertaining to *service outputs provided, channel efficiency and costs,* and *innovativeness and change* (9).

Marketing service outputs. The obvious measure of channel performance (its ability to deliver products to customers) is usually taken to be the volume of products delivered over a period of time. However, in addition to the physical delivery of a product, a channel provides various marketing services, or marketing outputs (7). Production and consumption acts are often performed at different locations and times and require diverse technologies. Producers and consumers may differ as to the assortments created or used, the sizes of the units involved, and the frequencies of production or usage. These differences imply the necessity of marketing activities that reduce discrepancies and thus create place, time, size, usage, and assortment utilities for the customer.

While several researchers have explored the issue of service outputs and their measurement, the results are far from being conclusive. Bucklin has specified four service outputs: (1) spatial convenience (or market decentralization), (2) lot size, (3) waiting or delivery time, and (4) product variety (assortment, breadth) (7). Clearly, however, the list can be easily extended. Availability of credit, quality maintenance, risk reduction, viability-stability of supply, realization of potential transactions, and availability of personal service and attention may also be added (9, 16, 19, 41, 49).

The type of outputs that are important may vary from one channel to another. As indicated by Figure 8.1, the relative importance of the outputs depends on the usage and production technologies involved and specific product characteristics.

Cost and efficiency. While the provision of service outputs by a marketing channel increases consumer utility, it also increases costs. Thus, the ability of a marketing channel to provide such service outputs is constrained by its efficiency and customers' willingness to incur pertinent costs.

Conceptually, channel costs may be thought of as the costs of providing delivery of the product to its customers and the costs associated with providing him with diverse marketing service outputs. The two are interrelated. Thus, one can establish some level of service, measure the costs associated with delivering products at that service level, and compare them with costs at higher levels of service.

Innovativeness. The ability of a channel to adjust quickly and properly to changes in demand—supply conditions, customer needs, and technological changes—is a major factor in determining its long-run survival. This is dependent on the openness and closedness of the system and its ability to absorb new as well as reject old and useless components. When the system is relatively closed with barriers to entry and centralized control over entry, its ability to respond and adjust appropriately to environmental changes may decline.

Allocation of Marketing Functions

Activities providing the service outputs desired by users often are performed at least partially by the users themselves. In that sense, we can view users as being part of the marketing channel. The degree of users' participation in such activities will affect the amount of work to be done by the commercial (nonuser) channel members. Thus, a process of adjustment takes place. Users must evaluate different mixes of service and determine the extent of their own involvement in each activity. The problem is that both direct participation and reliance on the commercial channel members to perform the desired activities involve certain costs. The cost associated with commercial members is perhaps more obvious. If those are relied upon to provide the pertinent services, the price charged by the retailers, wholesalers, and producers will increase. This truism has been formalized in the marketing thought as the "wheel of retailing" law which observed that trading up in retail institutions is associated with rising prices (26). Partial empirical evidence for the relationship between level of service and margins was also provided by Etgar (16).

The costs to users of self-performance of marketing activities may be less obvious. Such costs reflect expenditures made by users as well as the imputed values of users' resources invested in the performance of the pertinent activities (4). Industrial users, consumers, and scientists often devote substantial amounts of time to the performance of marketing activities. The value of that time may be considerable (17). Proper accounting for marketing costs associated with self-performance must then include the value of the time investments. As the degree of user participation increases, the greater are the costs that he incurs. But at the same time, the lower are the costs to the commercial channel members, which allows them (presuming no monopolistic trends develop) to offer products at lower prices to the user. Similarly, as the user reduces his direct participation, he needs to rely more on commercial channel members. While his direct costs may go down, his indirect costs (higher prices to channel members) will increase. The ideal allocation of tasks between users and the commercial channel members is one that minimizes total costs. The result of the process is the development of a set of arrangements that stipulate the activities to be performed by the customers as well as the specific tasks to be performed by the commercial channel members.

Channel Structure

The demand for specific channel tasks in terms of various marketing services generates a search for a pertinent channel structure. The concept pertains to arrangements of economic entities in terms of institutions, spatial settings, and allocations of decisions, activities, and communications among channel members. The channel structure that evolves should be so well adjusted to its tasks and environment that no other channel arrangements could be more attractive to users. That ideal structure is defined as "the normative channel" (7) and represents an equilibrium situation.

A question of major importance regarding channel structure is the role of channel intermediaries. Channels often vary in terms of their length, or the number of layers of intermediaries they include. While some channels exhibit few layers (short channels) others are composed of several layers of different intermediaries (long channels).

The design of a proper channel structure does not depend only on channel tasks selected. The channel designer must face a number of environmental constraints which may effectively limit his freedom of choice. In particular, *technology, legal environment, product characteristics,* and *patterns of competition* may affect channel structure by creating conditions that are more or less conducive to the development of specific structures.

Channel Structure and Performance

The potential relationships between various channel structures and performance indices are the major issues with which channel theory has to grapple. If such relationships can be established, channel managers can expect specific results and control impacts by designing or selecting appropriate channel structures for the distribution of their products.

A useful tool of analysis is the structure-conduct-performance paradigm developed in industrial organization theory (46) and recently applied in channels theory (9). The paradigm depicted in Figure 8.2 states that performance of a system (an industry or a channel) is the end result of a large set of decisions and activities made by managers operating in the channel. Their decisions and activities regarding pricing, location, market transactions, etc. (conduct), impinge directly on channel tasks performed and consequently upon customer utility.

The conduct of channel members is often influenced or constrained by the characteristics of the environment in which they operate and the structural characteristics of the channel system. The latter may determine the amount and degree of competition in the channel, its degree of openness and closeness, the ability of the various members to achieve their goals, and their need for cooperation.

Channel planners can attempt to affect channel performance in two ways. They can do it directly by designing policies for influencing the decision makers in the channel. Alternatively, they may try to design the channel structure in a way that is conducive to the achievement of present channel tasks. In many cases, channel structure manipulation is a more attractive strategy for the channel planner. He may find that his ability to influence channel members directly is limited or ineffective because their goals conflict with effective performance of channel tasks or because an existing channel structure is not capable of performing as required. Structure manipulation, on the other hand, allows the creation of suitable conditions required for desired performance.

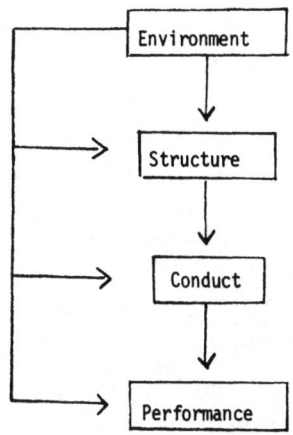

FIGURE 8.2: THE STRUCTURE-CONDUCT-PERFORMANCE PARADIGM

Channel structure and channel tasks. A major factor that may affect selection of channel structure is the extent of marketing tasks facing the channel. *In general, a greater demand for marketing tasks will increase the desired length of the channel by increasing the economic advantages of use of channel intermediaries.*

The economic rationale for the emergence of channel intermediaries is their ability to reduce channel costs. Several rationales have been proposed to explain economies stemming from the operations of channel intermediaries. Stigler (50) has demonstrated that emergence of intermediaries can provide a channel system with substantial vertical economies of scale. Specific functions can be "spun off" from producers or users to intermediaries who can specialize. Because the latter enjoy economies of scale, they can accomplish these functions at lower costs and reduce total system costs.

Another approach, developed by Balderston (1), emphasizes the role of a channel intermediary as a communicating clearing house who can reduce the number of contacts required to clear a given market with a specific number of transactions to be undertaken.

Balderston's model is simplistic. It assumes that a market where all sellers contact all buyers is replaced by one where all sellers and all buyers contact the middlemen but none contact each other. Baligh et al. relaxed some of these assumptions and concluded that there are substantial boundaries to the advantages of the middleman as a communication link (3).

A third rationale for the emergence of channel intermediaries developed by Baligh and Richardz (2) emphasizes their ability to reduce the negative effects of variations in supply and demand. The uneven conditions on either side of the market may create situations of long queues at supply or demand depots as buyers or sellers wait for overtaxed suppliers or users. At the same time, at other places, nonutilized capacity may stay immobilized due to a random decline of demand. By pooling supply and demand orders, intermediaries can smooth random variations at individual demand or supply stations, eliminate most of the queues and shortages, and reduce system costs.

Bucklin in his *postponement-speculation model* (6) suggests that channel intermediaries can substantially reduce the risks of operation for the channel unit as a whole by allowing it to delay risk-carrying activities such as bulk breaking until relatively late stages of channel operations.

Bucklin and Carman (9), Bucklin (8) and Etgar (16) suggest that channel intermediaries may become more economic when there is a greater demand for more extensive service outputs. When customers demand greater market decentralization, larger assortments, and more frequent purchases at smaller sizes, those may be more economically provided through longer channels. Thus, a greater market decentralization that allows customers to purchase products closer to their place of residence implies that the marketing channel should maintain a large number of retail outlets extending communication and physical distribution lines. This offers some economic opportunity for channel intermediaries who can reduce communication or delivery costs.

Channel structure and channel efficiency. Lengthening the channel through the employment of specialized channel intermediaries may increase costs and reduce efficiency. The additional costs may stem from two factors. First, they may reflect the additional expenditures associated with providing a higher level of service outputs that are characteristic of longer channels. Second, they may reflect inherent inefficiencies that are often built into longer channels.

The inefficiencies of longer channels are associated with the fact that such channels embody independent channel intermediaries who desire to maintain their independent status and that the increase in the number of channel levels increases costs of communication and intrachannel interaction.

The drive for independence may often induce channel members to control or participate in a large number of channel activities. As a result, duplication may be prevalent in the channel as different channel members perform similar tasks. For example, in the property and casualty insurance industry, insurers and insurance agents had performed similar billing and accounting procedures of maintaining and monitoring accounts receivable (15). For reasons of independence, channel members may also resent channel functional adjustments. Even though a specific function could be performed more efficiently elsewhere in the channel, current performers may resist such a change. Finally, in order to maintain their independent status in the channel, channel members may insist on maintaining diverse sources of supply, patterns of operation, and selling procedures and reject attempts for standardization. Fragmentation of channel operations leads however to additional channel costs and reduces the cost-saving opportunities embodied in the application of large-scale standardized equipment and procedures.

The increase in the number of channel levels may also delay flows in the channel. Information can often be distorted or omitted as it is conveyed through layers of channel members. Physical flows may be delayed by the necessary unloading, processing, and reloading in each station along the channel route.

Another source of growing costs is the increased cost of search activities in longer channels, which usually exhibit a larger number of channel establishments at the various channel levels. Producers, consumers, and channel intermediaries are not restricted to a specific subset of other channel members, but can interact with many. This freedom tends to create a continuing shift in channel alignments where trade partners are dropped and new ones picked up. Channel members have to devote their resources to the search for and negotiations with their trade partners.

Channel structure and innovativeness. The evidence regarding innovativeness in differently structured channels is mixed (9). Kriesberg (30) and Davidson (14) suggest that channel innovations are often introduced into channel systems by outsiders. Such firms are not bound by industry and channel traditions and are not committed to substantial investments in the old procedures and technologies.

The importance of outsiders for channel innovation suggests that channels that are relatively more open and exhibit lower barriers to entry will be more innovative. This may imply that longer channels are to be preferred. Such channels are more loosely managed and entry is easier at many channel levels. A short channel with fewer institutions and tighter controls may be less conducive to the entry of new firms.

However, Bucklin and Carman suggest that short, vertically integrated channels may have their own advantages regarding adoption of innovation (9). In such channels, power tends to gravitate to a channel institution that can use it to enforce change on resistant channel members. The channel "captain" can also recruit better channel financial resources and can harness them to invest in new technologies or in the production, distribution, and promotion of the pertinent product.

STI Channel Structure and Performance

Marketing Tasks in STI Channels

The first part of this paper suggests that key factors in designing a channel are the set of marketing services desired by customers, the pattern of allocation of marketing works between customers and channel members, resolving channel tasks, and the pertinent costs involved. In this section we shall review the relevance of these concepts to STI channels.

Marketing Services

Several factors may be suggested as potential services to be demanded by STI users.

Exchange effectiveness. A major role of a marketing channel is to achieve a maximum exchange effectiveness. This concept, developed from works of Sosnick (47) and Preston and Collins (40), suggests that in evaluating a channel, its ability to link together producers and users in meaningful transactions should be considered. Thus, if one channel system generated more exchanges than another, the former can be defined as being more "effective" (though not necessarily more efficient). In channels where market mechanisms regulate flows and exchanges, the number of transactions at ultimate markets is fairly approximated by volume of sales. In STI channels, ultimate market exchanges may be insufficient measures of performance because information exchange markets often do not exist for scientists. Instead, physical or other measures of flows may have to be utilized.

In the STI industry, the problem of exchange effectiveness is especially important in view of the enormous amount of scientific information available but not always utilized. An STI channel should, therefore, strive to maximize the number of exchanges that should take place in the channel.

Yet an attempt to define specific channel tasks along this dimension of service may generate problems for channel designers (12). First, it is not clear whether all transactions should be treated equally and to what extent exchanges with diverse market segments are of equal importance. Primary, secondary, and even tertiary market segments may be defined and different penetration goals may be set for each. In that case, some corresponding weighting system must be developed that will facilitate the comparison of various penetration mixes.

Second, the concept of exchange in STI merits discussion and analysis. The usage of scientific information will vary from one scholar to another and across situations. In some cases, STI will be used as input into a new research; in others as a confirmation for an on-going research, while in other cases, it may convince a scientist to forego a research idea. The varied uses raise the question of whether all should be considered as having equal value.

Traditionally, this issue has been circumvented by focusing on the demand for information of scientists irrespective of their motivations, and the capacity of the STI channel to satisfy such demands. This approach raises yet another set of issues. It assigns to the STI channel the passive role of *information supplier* and not of *information marketer*. The difference is substantial. The former responds to a given call for information; the latter attempts to identify customer needs and provide by itself the information packages responding to these needs.

Thus to measure the exchange effectiveness of a particular channel from a marketing point of view there is a need to develop measures of consumer needs and of ways to satisfy them.

Fast channel flows. Scientists often must rely on recent publications for their work. Excessive delays between the time research takes place and publication of its results may reduce the usability of a specific manuscript. Fast transmission of research results to scholars helps scholars resolve issues in their own research and saves them from the dangers of working on topics already discussed or problems already solved. Finally, the researcher's image as a quality scholar is often dependent on his show of up-to-date knowledge about the researched topic.

Proper assortment. The ability to get all the information pertaining to a topic is of primary importance to the researcher. The importance of this factor has increased lately with the increased emphasis on interdisciplinary research. In his core discipline or research area, a researcher is often well attuned to the work done by his various colleagues through personal contacts, drawing on literature, preprint circulation, etc. He may not be so well informed as to the status of research in fringe areas which may be important to his research.

Appropriate information formats. A researcher may be very interested in receiving information in different formats at the different stages of his research process (22). (See Figure 8.3.) In their study of reading habits of scientists, Martin and Ackoff have found that scientists spend close to 60% of their reading time reading articles, 5-7% reading digests or abstracts, 4-5% reading tables of contents, and 1-2% reading reviews (34). Thus a scientist may be concerned with the variety of information formats received. Ideally, he would like the distributive channel to be capable of feeding him with proper information as he

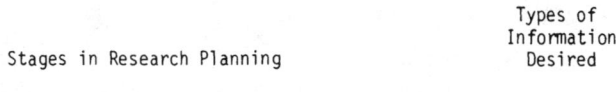

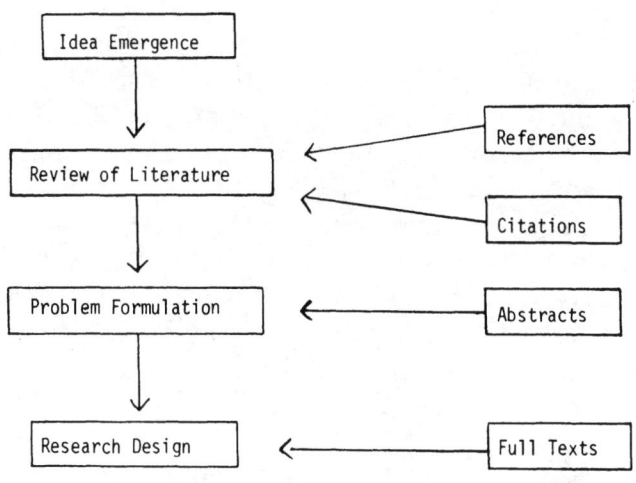

FIGURE 8.3: STAGES IN RESEARCH PLANNING AND
PERTINENT INFORMATION REQUIRED

moves through his research process.

Grading and classification. The amount of information facing the individual researcher is often so vast that he cannot possibly review it himself. As a result, researchers need classification and grading services that will provide them with prior indicators as to the content, quality, and relevance of journal articles and books. Thus, in this respect, users of STI do not differ much from buyers of commodities such as cotton, coffee, or iron ore who have similar needs for grading and classification.

Coordination of demand and supply. A major task of the marketing channel is coordination of demand for and supply of the products or services it carries. The coordination between supply and demand is often achieved in markets with the assistance of price and related market mechanisms. As prices rise or fall, producers are notified about the relative desirability of various products and increase or decrease production appropriately. In centrally coordinated channels (administrative or contractual) where intermediary markets play only a minor or no part at all, information about customer preferences are transmitted directly to producers through specially designed communication systems.

In STI channels, the need for demand and supply coordination is particularly acute since, traditionally, these channels did not use intermediary market mechanisms. This does not mean that STI producers do not face or respond to demand for scientific information.

Technological innovativeness and institutional change. Another major index of performance is the ability and speed of an STI channel institution to adopt technological innovations. The inability of some institutions to accept new information storage retrieval and transmission technologies may be a substantial obstacle to the achievement of greater channel efficiency.

Technological change is often coupled with institutional change. Existing channel institutions may become obsolete due to changes in customer patronage habits and usage patterns or because of technological change. New channel institutions may need to be developed to respond to the new needs and opportunities.

Costs and efficiency. In STI, channel efficiency refers to the costs associated with transmission, storage, and retrieval of a given bit of information. It is important to recognize that the costs of distribution of STI include not only the costs borne by the various channel intermediaries but also those incurred by customers and producers.

Thus, producers often have to make great efforts to convey their manuscripts, to communicate with journal editors, and to edit and correct manuscripts according to editorial requests and reviewers' comments. All this demands a substantial investment of time and other resources (money, institutional funds, writing material, assistants and secretarial help, etc.) from the producers. These expenditures must be added to the overall costs of each channel.

Allocation of Marketing Services Between Customers and Channel Members

Users often perform some of the pertinent marketing functions in the STI channels especially in regard to the search for and storage of STI. They spend a substantial amount of their time in searching for articles, following citation leads, and using their time, private index systems, secretaries, and research assistants to create individual information storage and retrieval systems. According to Martin and Ackoff, between 50 to 60 percent of their reading time (depending on the discipline) is used by scientists for nondirected browsing, i.e., on search efforts rather than on specific scientific reading (34).

The costs of search activities are high. Scholars must divert time and research resources from idea generation, information processing, and analysis to information collection and preparation activities, thereby perhaps reducing their effectiveness as researchers. Furthermore, in many cases an individual researcher cannot perform specific functions. All of this implies that customers-researchers may desire to shift a large share of the search activities to channel intermediaries.

STI Channel Structure

In Figure 8.4, four different STI channels that are frequently used are presented. Each will be analyzed separately.

Zero-level channel. On the left side of Figure 8.4, a direct, or zero-level, channel is depicted. Here, producers of STI convey it directly to customers (other researchers-users). The importance of this type of a channel for the dissemination of STI has been expounded by Garvey et al. in their series of studies of STI diffusion patterns among scientists (20, 21, 22) and confirmed in several other studies (10, 24, 29, 43, 44). Scientists utilize personal correspondence, telephone discourses, national and regional meetings, and visits to communicate on a personal basis.

The major advantage of this type of a channel is the *speed* at which scientific information is transmitted from producers to consumers. Such a channel also offers the producer an oppor-

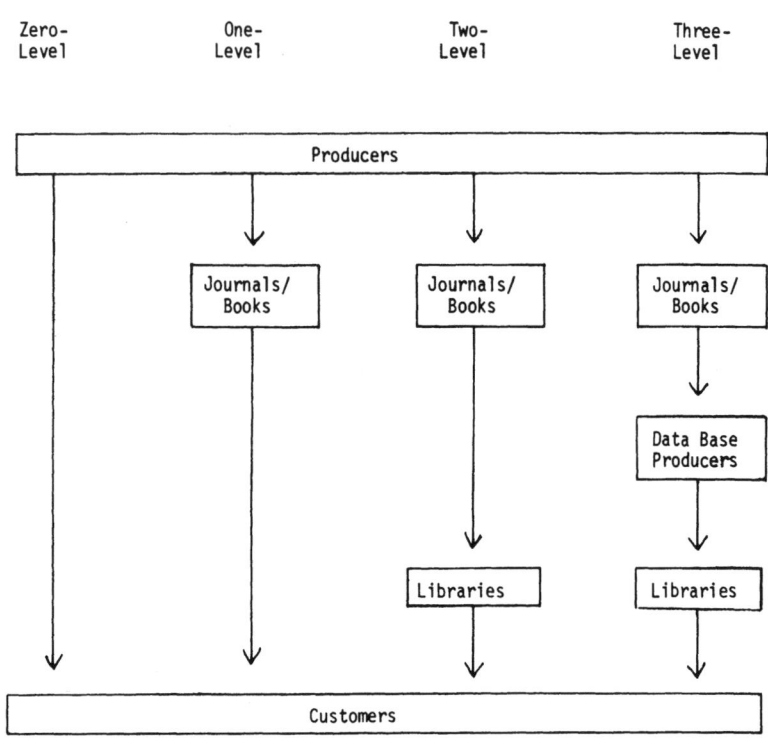

FIGURE 8.4: CURRENT STI CHANNEL TYPES

tunity for direct feedback including critiques of his work and information concerning topics explored by his colleagues, how up-to-date he is in his research, etc.

Most of the zero-level channels are restricted to a small group of researchers who know each other's work amd who form "informal colleges" (42). This sociological structure may involve some costs, since the channel cannot provide effective quality control mechanisms or check for duplication in effort. This imposes some burden on customers who have to rely on "Brand names" of the scholars to serve as quality indicators.

The informal modes of interaction in such channels may make it very difficult for producers or customers who are unknown to the "informal college" members to penetrate the channel in order to disseminate their material or be exposed to others' work. This may reduce the applicability of the work disseminated in this channel and reduce its market coverage. Also, it may tend to block the dissemination of relatively new and revolutionary ideas which are often introduced by young scholars and scholars from the fringe of the discipline.

One-level channel. A popular channel of distribution for STI is a one-level channel in which producers submit manuscripts to journals, books, and other types of published media from which they are conveyed to the customers. The use of printed media has several advantages to the customer scientists.

First, journals can provide scholars with information about activities of other members of the discipline, thereby providing a larger *variety (assortment)* of information than is available through a zero-level channel. Second, journals provide customers with greater *location convenience* as they are delivered at the scholar's home or office and do not require any travel. Third, they also provide *"quality control"* service. Most journals have complex acceptance procedures designed to ensure certain standards. As journals vary in their quality control requirements, consumers may establish ranking of journals by levels of quality and use the locus of publication as an indicator of the quality of the published manuscript. Explicit reviews, especially of books, provide additional quality control information. The quality control performed by the journals greatly relieves the scientist-user of the need to exercise his own quality evaluations and reduces the risk associated with using low quality inputs into his own work.

Fourth, the journals provide users with some indication as to the *structure of demand* for scientific information. By following up publication in journals, a scientist can learn which topics are of current interest to reviewers. As reviewers are usually senior scientists, well acquainted with the field, their interests may reflect interests of the discipline.

At the same time, the existence of a large number of journals catering to diverse disciplines often provides ample opportunity for the *expression of diverse opinions* in scholarly works. When a new topic becomes important, an appropriate journal can be established to become a tool of communication among interested parties. This flexibility ensures that some freedom in terms of publication can be achieved. Thus, scholars with diverse or revolutionary opinions (e.g., Marxist economists) may begin to publish their own journal.

The one-level channel has several major disadvantages. Its major disadvantage, as compared to the zero-level channel, is the *long delay* before a paper is published (54). Papers have to be submitted, screened by editors, and reviewed. To be reviewed, submitted paper have to be read by several referees; if accepted, the paper is usually returned to the writer for corrections or clarifications and may be reviewed again. Then it is scheduled for publication. As most journals experience some backlogs, publication can take many months or years. In many cases, manuscripts are rejected by the first journal to which they are submitted and have to be resubmitted elsewhere, thereby substantially lengthening the time span involved. Scientists who rely on journals for supply of scientific information may often find that the published information is already outdated by the time of publication.

The fragmentation of literature into specialized journals is a natural response to the problem of an expanding literature. Yet it brings problems too. Readers are faced with an increasing number of journals to scan. The same topic may be covered in several journals and one author may disperse his work in various journals.

Fragmentation brings problems for authors as well as for readers. Given an ever-increasing number of journals whose scope and editorial policy may not be easy to ascertain, authors may be faced with considerable problems in deciding where to submit their work.

The existence of a large number of journals with their separate editorial and publishing staff increases the overal costs of STI distribution and builds a substantial amount of *excess capacity* into the system. People who could edit, review, or publish a larger number of manuscripts are constrained to work with the smaller subset of manuscripts submitted to their own specific journal.

Similarly, the existence of many journals in the same or similar fields generates *duplication* in the reviewing process and additional storage costs to customers and libraires. Because of high rejection rates of many journals (54), some manuscripts are submitted sequentially to several different journals and must be reviewed twice or more. Multiplicity of journals also forces customers to stock more volumes and use *more storage space* thus increasing costs.

Storage costs are especially important because of the indivisibility of the "product." In most cases, articles published in each journal edition may vary in demand. Only a few generate substantial interest, while others may be of no interest at all. Yet the customer is forced to buy and stock the whole volume.

Two-level channel. In the two-level channel, another type of intermediary institution appears: the library. Published journals are not delivered directly to customers, but shipped to libraries. A two-level channel is thus created in which the library performs the role of an *information retailing outlet.* It offers customers the availability of a large number of publications at one location and usually at a relatively low cost to the customer. The customer-researcher can gain access to the material in the library by borrowing it, using it on premises, or duplicating it.

The importance of the library as a supplier of scientific information is highlighted by Martin and Ackoff's study (34). The research found that chemists do over 50% of their scientific readings in libraries while physicists devote 33% of their reading time to that purpose (34, 33).

While library use increases the assortment of conveniences for the user, it often also *lengthens the delivery time.* Information has to be shipped to a new station in the channel; this implies additional time delays. Once in the library, the journals must be processed, catalogued, etc., again lengthening the period of time during which they cannot be used by readers.

Weinberg argues that in spite of their impressive growth in terms of volumes handled, library service has declined in the past few decades (52) primarily because of three factors. The increases in the number of sources of information (books, journals, periodicals, government publications, etc.) and of users have overwhelmed library resources. Second, due to the increasing specialization, researchers have developed increasingly narrower ranges of interests so that the number of works that can be shared with other researchers declines. A library thereby needs more volumes to provide the same amount of information per user. Finally, the growth of interdisciplinary fields has made it difficult to identify a book or a journal with a particular user. Thus, a researcher may have to search in many libraries before finding the information he needs.

A library may also require substantial *user search costs*. For historical reasons, libraries currently conduct their operations as if their strategy were to store and provide access to books, periodicals, papers, and other materials. Weinberg argues that libraries' emphasis is on storage to the detriment of their ability to provide access; and the example that he gives is library binding policies. At the end of each year, the library gathers together all of the issues and sends them to the bindery. This is done to prevent the loss of single issues which are often difficult to replace and protect. Weinberg argues that it is during the period just after the issue appears that a journal is most in demand. Thus journals are taken out of circulation in their most demanded period to allow for use when they will not be heavily demanded.

The use of a library also reduces the *spatial convenience* of the customer. It forces him to leave his office where his notes or research materials are located and go to the library (32). Finally, the use of a library substantially *increases system costs*. In the zero-level system, each customer maintains his own STI inventory; retrieval is usually simple and quick. The situation is different in a library. Because of the fragmentation of the scientific literature and the emergence of hundreds of diverse journals, a library has to maintain a large volume of periodicals to ensure delivery of a reasonable level of scientific information to its customers. Because many journals are read by scientists from different fields, duplication may often be required. The same journals must be stocked in several libraries that serve different constituencies. Often, this becomes infeasible and libraries must revert to complex mechanisms of interlibrary specializations and transfer of material. In most libraries, many costs are treated as overhead and services are priced on a marginal basis to attract users.

Since individuals request information while organizations often pay the cost, information may be virtually free to the user. At the margin, this certainly causes *inefficient use of information*. Customers use information up to the point where its value equals the costs of information use for them (i.e., value of their time and resources) without considering the overhead costs. As a result, information may be used over and above its real marginal costs.

Three-level channel. A three-level channel also hosts an information intermediary in the form of various secondary services, such as data base providers. Such an institution provides references to and frequently abstracts of portions of the total mass of literature, often in an indexed or classified form. The portions chosen can be either topic or discipline-oriented like *Chemical Abstracts*, mission-oriented like *Food Science and Technology Abstracts*, quality-oriented like the *Science Citation Index*, or with a national orientation like *British Technology Index*. Their coverage policy can be selective, usually on a quality basis, or comprehensive, and the output is generally available in several formats, such as cards or magnetic tape (39).

While the total number of data base providers is uncertain, their numbers are growing fast (39). In 1963, the National Federation of Science Abstracting and Indexing Services was able to identify close to 2,000 entries and since then it has increased even more.

In marketing terms, a major role of a data base provider is to generate for the customer information about an *assortment of STI* while sharply reducing the amount of his search effort involved. Instead of searching through the libraries and tracing citations from one article to another, the customer receives all pertinent information about STI in a packaged and organized form.

Data base providers also perform the pertinent function of *bulk breaking* in information processing, namely reducing scientific information reports to smaller, more digestable units that can be utilized by scientists in the various stages of their research. The bulk breaking allows researchers to learn superficially about research and determine whether it is of interest to him or not. If it is, the researcher will retrieve the full article. In that way, scientists' direct search costs are further reduced.

The introduction of a data-base provider usually increases

overall costs to the channel. At the same time, the use of data bases does not reduce many of the other costs of an STI channel stemming from excess capacity, high storage costs and duplication. Furthermore, lengthening the channel *increases delivery time*, as information has to be moved to another station in the channel.

The Emergence of A New STI Channel

The pressures for more marketing services, less customer involvement, and lower costs create a constant driving force toward changes in STI channel structures. Such changes focus on the emergence of a new type of channel intermediaries—information brokers. Recent advances in computer technology have resulted in the development of highly efficient and fast methods of storing and retrieving information that can be used for the dissemination of STI.

At the heart of the new channel system is a new type of STI intermediary—a computer-based bibliographic information storage and retrieval system. In such a system, the contents of articles of many journals are analyzed. For example, the MEDLARS system at the U.S. National Library of Medicine produces an index for the contents of 2,300 journals (13). The DRUGDOC system covers 3,400 major journals in the biomedical field (45). The contents are indexed based on a controlled vocabulary.

In indexing, special effort is made to ensure that a specific article will appear under as many vocabulary labels as possible. This is to ensure access to the article to readers with diverse interests. A computerized thesaurus of all usable terms is thus created and articles are assigned to the various terms. An information seeker interested in a specific term will thus receive a list of articles that pertain to the specific term and were scanned by the system. Alternatively, a user may prepare a profile of research interests. The profile will be deposited in the computer's memory. Each month it will be matched with the data base and a printout of relevant titles, abstracts or citations will be sent to the user.

Information brokerage services of that kind are offered by several commercial firms, by regional information units, data source providers and by large university libraries (45). Scientists who wish to use these services usually must do so through a business, university, or research library that subscribes to the services.

Potential New Channel Structures

The relatively high costs of computer technology today still limit the role and opportunities of the new STI intermediary. As the price of computer systems, hookups, and terminals goes down, a more expanded role for this new intermediary can be envisioned.

The potential of computerized services to store and retrieve information may lead to an expanded role for these intermediaries well beyond storage and retrieval of indices, abstracts, or names of articles. Low cost storage, retrieval, and information manipulation may completely revolutionize STI distribution. At the same time, the external pressures may contribute to a development of new types of channel intermediaries.

In the following section, several scenarios have been developed describing potential new channel structures. Any of these scenarios could take place in the near future.

A major trend that may emerge in the STI channels as a result of the emergence of the new intermediaries is a movement toward vertical integration. Vertical integration serves to reduce some of the costs inherent in existing channel structures and to increase channel efficiency. The direction of the vertical integration could be well toward both ends of the channel.

Forward Vertical Integration—Information Center As A Library

One possibility is that as the memory capacity of a computerized service increases and costs of using it decline, computerized data-processing information centers could reduce the role of the library as a scientific information retail outlet. In such a channel, all published data—journals, books, government publication, etc.—would be directly fed into a computer memory to be retrieved by users (Channel Type A in Figure 8.5).

A caller interested in a specific article or publication would

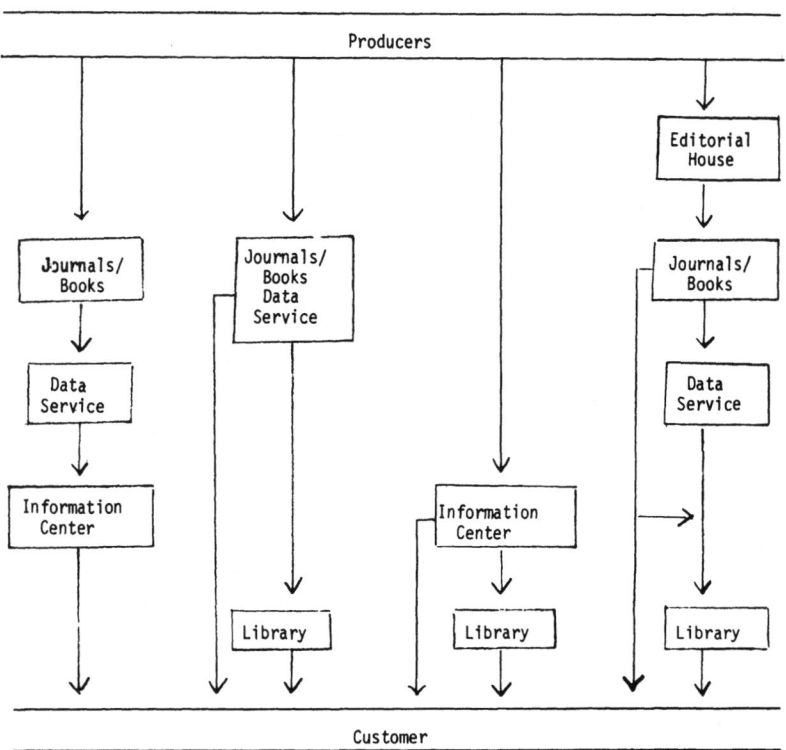

FIGURE 8.5: FUTURE POSSIBILITIES FOR STI CHANNELS

be able to request not only its title or abstract, but also the article itself to be printed out or projected on a TV screen. A service that indicates development in this direction is the Mathematical Offering Service (MOS) offered by the American Mathematical Society (31). This service offers reprints of articles that satisfy the criteria specified in a customer's interest profile. In addition, the subscriber indicates authors whose work he specifically wishes to receive or exclude.

As a result of such services, *libraries may become less important.* While some may become, or integrate with, such computerized centers, others may disappear once the business research or educational institutions maintaining them decide that it is more economic for their needs to subscribe to a computer-ized service instead of maintaining an independent library. Librarians would be relieved of the need to assist users in searching for particular publications, since it could be more efficiently done on the computer. Librarians as we know them today would disappear, to be replaced by computer program-mers and technical experts. In the place of a library with many stacks of books there would be computer input-output hookups into which customers type their requests and from which they receive information.

At the extreme case, *the physical concept of a library may also disappear.* Users would be directly hooked up to the com-puterized system via input-output installations in their work-places. The users would thus be able to interact directly with the computerized data processor without an intermediary.

A major advantage of this new variety of channel is its poten-tial to reduce the storage and retrieval costs that plague today's libraries. Several libraries might be replaced by one information center, thus eliminating needs for separate (and often duplicate) storage of STI, cataloguing and bibliographical work, and several complex retrieval systems. A computerized system would also reduce the costs of information retrieval by allowing a greater degree of bulk breaking. Users would be able to receive a copy of a specific article in a journal or a chapter in a book without having to retrieve the whole volume, or they may receive an entire collection of pertinent STI materials.

Computerization of library stocks would also *speed up* the retrieval process. If the material is stocked by a particular center, it could be immediately printed out for the reader. How-ever, it is sensible to assume that some specialization will take place wherein different information centers will store different types of information. In that case, a user requesting a material that is not available at his information center would have to wait till it is searched for at other centers and then rerouted to his center. Because of the high speed of electronic communica-tions, such search processes can be expected to be much faster than interlibrary loan procedures used today.

Another substantial advantage that such information centers will have is their ability to *monitor requests* (whether filled or unfilled) *for scientific information*. The records of such requests can be used to measure interest in a topic, thus replacing various indirect measures (23). A more accurate and timely knowledge about user interests may assist producers in focusing their research on more relevant or popular topics and issues, thus increasing the effectiveness of STI production and distribution.

The development of such information centers and their replacement of libraries may also contribute to the *development of information markets* by revolutionizing approaches to cost allocation of information use. When subscribing to the new information center services, some institutions may elect to maintain the traditional approach. They will bear all of the relevant costs and maintain the no-cost approach for individual users. Other institutions may attempt to introduce marginal cost pricing for the individual users. This may economize use of information. The users will compare their prior expectations of value of the additional knowledge to be derived from a proposed piece of information with the associated costs and decide accordingly. A more economic use of information may emerge. Less frequently used items may be dropped from the stocks of regular information centers and delegated to a special subsidiary file, while the more popular information pieces will become more readily accessible.

However, in spite of these advantages, such vertically integrated systems may have substantial disadvantages in terms of *loss of customers' control, personal services, and flexibility*. Users who will replace their libraries by a subscription service will have to realize that to a large extent they may forego some or all control over stocking decisions, since these will become the responsibility of the managers of the centers. While the range of available choice may be expanded by tie-ins to other information centers, the chances are that off-beat, new revolutionary ideas and information sources will not be stocked. Information center managers may emphasize the stocking of the more popular publications which are more economic (i.e., which have a higher turnover) and may be reluctant to stock less-used manuscripts. As a result, it might be more difficult to convey new ideas and research in such a system.

The replacement of the personalized services of a librarian by those of a computerized program may be seen also as a disadvantage by many users. Individually tailored assistance would be replaced by programmed service. In spite of its advantages the latter is very inflexible and cumbersome to change. Also the partial or complete disappearance of open stocks implies that a major scholarly activity, i.e., undirected browsing, may become more difficult. This may adversely affect the process of scholarly creativity, which sometimes is based on inductive cognitive processes of distilling, analyzing, and adjusting several seemingly unrelated facts, research results, and concepts.

Finally, it should be remembered that such a system forces users to substitute machine interactive search effort for their personal search efforts. The advantage here is clear if the latter is less expensive than the former, but often this is be so. Some users may attach low marginal costs to their search efforts even if their average search costs are high. With information centers carrying out the search, economic price tags might be associated with each search. Researchers may be then less willing to search for marginally beneficiary information (32).

Forward Vertical Integration—Journals
As Secondary Data Sources

In this channel, some of the services currently available at secondary levels will be incorporated at the journal level in order to afford users greater selectivity in access to the content of journal literature (25, 31, 37, 53). One possibility is the two-edition journal concept discussed by Moore (31). Under this concept, a journal is issued in two distinct editions. One edition, directed to the broad audience, contains only the material of central interest. These may include descriptions of the topics discussed in each article—hypotheses, methodology, main findings, etc. The other, directed for a limited circulation, i.e., libraries, will contain the complete papers. Alternatively, journals may attempt to present compact units of information in the form of letters and communications that will briefly describe experiments, results, etc., and rely on the authors to supply full details to interested parties who will contact him directly.

The system offers several advantages. It can substantially

shorten the time lag before processed information (abstracts) is offered to customers by eliminating to some extent the need for secondary data sources. Customers can quickly scan through journal literature in search of relevant information without having to read whole articles. Thus costs will decline also. Users who are interested in processed information could pay only for that information and not for the complete journal. Journal prices might be lower and customer storage costs might also decline.

Backward Vertical Integration—Into Publishing

Another interesting scenario is the replacement of the scientific journals by a computerized service (Channel Type C in Figure 8.5). Papers will be put into machine-readable form and all subsequent editorial processing will be within the computer. In turn, users would have remote access to the content of the system.

Users interested in a specfic area of research will become aware of the new piece of work by scanning updated indices. Such a system will reduce the long delays that take place between the time of production of STI and the time it reaches the customers. In this sytem, the research data will enter storage soon after it has been generated and thus will become available to interested researchers almost immediately. However, such a system may lack *proper incentives to producers and quality control mechanisms.*

The use of published media such as journals provides a producer with two major advantages. The publication confers status and established property rights to the specific idea or results. In the proposed system these advantages may not be automatically available. Special arrangements will have to be made to ensure producers' ownership rights to publications.

The problem of a producer's professional image can be aggravated by the destruction of the "quality controls" built into today's journal publication system through the reviewing procedures. They ensure customers of publication quality and give a stamp of approval to the works of the producers. If the reviewing process is removed, any kind of information can be added to the computer memory without quality control. New

evaluation procedures may then have to be developed by customers of producers at great costs. Also, some information producers may be reluctant to contribute their work to the system,
seeking instead more prestigious channels where "ownership
rights" can be retained (such as the personal communication
channels).

It is feasible that an information center would attempt to
respond to the need for quality control by establishing its own
reviewal procedures. This, however, may endanger a major
advantage of the new channel, namely, speedy diffusion of
STI. Furthermore, the establishment of such reviewing procedures may seriously *harm freedom of entry.* The information
center would replace a large number of journals; thus the
number of entry points into the system would be reduced and
would presumably be controlled by the establishment of the
specific discipline. This may impose serious barriers to entry for
new STI products, organizations, and technologies. Thus, off-
beat ideas or controversial articles may be rejected without
having the recourse of applying to other journals or opening a
new one. New organizations could not enter the channel unless
accepted by the information center, which becomes an information gatekeeper. While channel conflicts may decline, innovativeness may be reduced as well.

Backward Vertical Integration—Into STI Production

The fourth scenario is one in which an STI channel intermediary undertakes to perform the activities that have been
defined as belonging to the STI producers (Channel Type D in
Figure 8.5).

A major service that could be provided by such an intermediary is matching of researchers. The information broker
could develop a mailing list of all researchers interested in a particular topic. A scientist who writes or plans to write a paper
about this topic could share his work or his thoughts with his
colleagues. Consider a scholar who has already developed an
idea or established a finding that he thinks is significant. If the
topic of his contribution is so specialized that there are only a
few such colleagues and potential users, the researcher can
reach them through the established lines of communication—

report, working paper, specialized journal. However, if he thinks that the topic is of sufficient interest to a large number of scholars he may be interested in the proposed matching service either to communicate with them quickly or because this may be the most effective way of reaching them.

A more complex version of this service performs higher-level tasks. One example is a counseling-article-writing service (28). In this case, the writer presents to a counselor the gist of his planned paper. The counselor uses a dialogue to help the author articulate his plans for the paper. He tries to represent the potential referees, readers, editors, and users of the proposed paper and their points of view, and may also assist the writer in selecting the appropriate journal to which the article should be sent.

Additional higher levels of services can also be envisaged. The researcher may agree to delegate completely the publication activities of his papers to middlemen. In this case, the role of the researcher will end with the writing of the article. The middleman will select the appropriate journal, submit the manuscript, and if necessary resubmit it to another journal. He may pool referees' comments, advise the author how to revise the manuscript or even do it himself and resubmit the article.

Undoubtedly, such services may be expensive. Yet, to young scholars about to enter a field, they may be well worth the price. Experienced authors, too, would find the service helpful. It will provide them with up-to-date information about journals in which to place their own publications.

Such a service may be particualrly valuable in view of the growth in importance of new user groups and media outside the United States, and the increased trend toward interdisciplinary research which is often directed toward people with whom a particular author may not be acquainted. The number of new journals is substantial and there may be journals for which an article might be ideal but that the author does not know well.

A major noneconomic obstacle for the development of such a counseling service is sociological in nature. The norms of the scientific community may be in direct conflict with operations of such counselors. One could hypothesize that the use of commercial inputs in the writing, revision, or even publication search activities by any author may seriously harm his image as a scholar. If this is so, such attitudes have to be changed before such commercial entities could succeed.

Increased Role of Information Brokers in STI Channels

The increased capacity of information brokers to learn about user preferences and needs, coupled with their extensive capacities for storing and processing information, may propel them into a dominant role in the STI channel.

Because of their size these channel intermediaries may be few. Their ability to monopolize gatekeeping positions near the producer or customer end may increase their channel control potential. In that case, *STI channels, which today are fragmented into a large number of various small and medium-sized institutions where no clear power structure exists, may transform themselves into a short, tightly controlled system dominated by the new channel intermediaries.*

These new channel "captains" may use their power beneficially for the channel. Their size, resources, and monopolistic positions may allow them to invest capital in new technologies and organize the channel to adopt and react to environmental changes.

The information brokers might also use their capacities to evaluate demand in order to establish scientific preferences and unsatisfied scientific needs and then attempt to respond to it by commissioning corresponding research.

Summary

STI channels have been undergoing a constant process of channel lengthening, a process that has substantially contributed to customer welfare. It offers customers a greater assortment at retail level and has substantially reduced search costs. The journal institution also provides the customer with the important service of quality control and grading, while the existing diversity of journals, literature, secondary sources, and information centers ensures a greater flexibility of the system as a whole and relatively low barriers to entry for new manuscripts, STI institutions, and technologies.

However, long channels also have substantial disadvantages. The costs are high, especially in terms of storage of information, fragmentation of decision making, communication delays, noise, duplications, and high producer search efforts.

New computer technology and organizational structures could offset the disadvantages by initiating a trend toward vertical integration and shorter channels. Existing channel intermediaries—journals, libraries, and secondary sources—could be replaced by computerized data-processing or editing houses which will become channel leaders.

Major advantages of vertical integration up or down the channel are associated with the ability to reduce handling costs and speed communications and channel flows. Such channels would be more capable of reacting to demand requirements, providing information in a variety of formats and speeding up STI flows. They may reduce customer search costs, increase assortment, and increase channel efficiency.

However, such new channels may demand the sacrifice of some service outputs currently demanded. In information centers, personal service may be replaced by computerized programs. Freedom of decision making in the individual researcher may be reduced and some spatial convenience may have to be sacrificed.

Most important is that such structures carry with them a substantial danger. The centralization of decision making and monopolization of gatekeeping positions by a few institutions may reduce the freedom of entry into the industry of new STI products, organizations, and technologies and thus reduce the utility of the STI channels to operate optimally in the long run.

The emergence of any specific channel structure is far from assured. These new ones that may emerge will depend on such factors as the evaluations of the performance of different channel systems by customers, the economics of the various technologies, and social barriers to change.

References

1. Balderston, Frederick E. "Communication Networks in Intermediate

Markets." *Management Science*, January 1968, pp. 154-171.

2. Baligh, Helmy H. and Leon Richardz. *Vertical Market Structures*, Chap. 2. Boston: Allyn and Bacon, 1967.

3. ——, Daniel A. Graham, Eric Weintraub, and Morris Weisfeld. "The Transactions Middleman is Inefficient." Durham: Duke University, School of Business, 1973.

4. Becker, G. "A Theory of The Allocation of Time." *The Economic Journal*, September 1965, pp. 493-517.

5. Bowersox, Donald J. and E. Jerome McCarthy. "Strategic Development of Planned Vertical Marketing Systems." In Louis P. Bucklin, ed., *Vertical Marketing Systems*, pp. 52-74. Glenview: Scott, Foresman and Co., 1970.

6. Bucklin, Louis P. "Postponement, Specialization and the Structure of Distributive Channels." *Journal of Marketing Research*, February 1965, pp. 26-31.

7. ——. *A Theory of Distribution Channel Structure*. Berkeley: Univeristy of California, Institute of Business and Economic Research, 1966.

8. ——. *Competition and Evolution in the Distributive Trades*. Englewood Cliffs: Prentice-Hall, 1972.

9. —— and James M. Carman. *Vertical Market Structure Theory and the Health Care Delivery System: An Analysis*. Berkeley: University of California, Institute of Business and Economic Research, 1972.

10. Chen, Ching-Chih. "How Do Scientists Meet Their Information Needs?" *Special Libraries*, July 1974, pp. 272-280.

11. "Computerized Library Service Available." *The CBA Exiter*. Austin: The University of Texas 1975-76 Summer Session Issue, p. 4.

12. Cooper, W. "A Definition of Relevance for Information Retrieval." *Information Storage and Retrieval*, June 1971, pp. 19-37.

13. Corning, Mary E. "The U.S. National Library of Medicine and International Medlars Cooperation." *Information Storage and Retrieval* 8 (1972):255-264.

14. Davidson, William R. "Changes in Distributive Institutions." *Journal of Marketing*, January 1970, pp. 7-10.

15. Etgar, Michael. "The Effects of Administrative Control on The Efficiency of Vertical Marketing Systems." *Journal of Marketing Research*, February 1976.

16. ——. "Differential Strategies in Interchannel Competition." *Journal of Business Administration*, forthcoming.

17. ——. "The Household As A Production Function." In J. Sheth, ed., *Research in Marketing*. Vol. 1. Greenwich: JAI Press, forthcoming.

18. Fallert, Richard F. "A Survey of Central Milk Programs in Midwestern Food Chains." Washington: Department of Agriculture, Marketing

Research Report No. 944, 1971.

19. Fulop, Christian. *Competition for Consumers.* London: Allen and Unwin, 1964.

20. Garvey, William D., Nan Lin and Carnot E. Nelson. "Communication in The Physical and The Social Sciences." *Science*, December 1970, pp. 1166-1173.

21. ———. "Description of Machine-Readable Data Bank on Communication Behavior of Scientists and Technologists." *Selected Documents in Psychology*, 1972, pp. 2-3.

22. Garvey, William D.; Nan Lin; and Kazuo Tomita. "Research Studies in Scientific Communications: IV, The Continuity of Dissemination of Information by Productive Scientists." *Information Storage and Retrieval*, December 1972, pp. 265-276.

23. Garfield, E. "Citation Analysis as A Tool in Journal Evaluation." *Science*, 1972, pp. 471-479.

24. Havelock, Ronald G., ed. *Planning for Innovation Through Dissemination and Utilization of Knoweldge.* Ann Arbor: University of Michigan, Center for Research on Utilization of Scientific Information, 1969.

25. Herschman, Arthur. "The Primary Journal: Past, Present, and Future." *Journal of Chemical Documentation* 10:1 (February 1970):37-42.

26. Hollander, Stanley C. "The Wheel of Retailing." *Journal of Marketing* 24 (July 1960):37-38.

27. Kaven, William H. "Channels of Distribution in the Hotel Industry." In John Rathmell, ed., *Marketing in The Service Sector.* Cambridge: Winthrop Publishers, 1974.

28. Kochen, M. and R. Tagliacozzo. "Matching Authors and Readers of Scientific Papers." *Information Storage and Retrieval* 10 (1974):197-210.

29. Korfhage, R. R. "Information Communication of Scientific Information." *Journal of American Society for Information Sciences*, January 1974, pp. 25-32.

30. Kriesberg, Louis. "Occupational Controls Among Steel Distributors." In Louis W. Stern, ed., *Distribution Channels: Behavioral Dimensions.* Boston: Houghton-Mifflin, 1969.

31. Kuney, Joseph H. "New Developments in Primary Journal Publication." *Journal of Chemical Documentation* 10:1 (February 1970): 42-45.

32. Maier, John M. "The Scientist Versus Machine Search Services: We are the Missing Link." *Special Libraries* 65:1 (April 1974):180-188.

33. Martyn, J. "Secondary Services and The Rising Tide of Paper." *Library Trends*, July 1973, pp. 9-17.

34. Martin, Miles W. and Russell L. Ackoff. "The Dissemination and Use of Recorded Scientific Information." *Management Science*, January

1963, pp. 322-336.

35. Mattson, Lars. *Integration and Efficiency in Marketing Systems.* Stockholm: Stockholm School of Economics, 1969.

36. Mighell, Ronald and Lawrence A. Jones. *Vertical Coordination in Agriculture.* Washington: U.S. Department of Agriculture, 1963.

37. Moore, James A. "An Inquiry On New Forms of Primary Publication." *Journal of Chemical Documentation* 12 : 2 (1972):75-78.

38. Naert, Philippe A. and Alain V. Baltez. "A Model of A Distribution Network Aggregate Performance." *Management Science* 21:10 (June 1975):1102.

39. National Federation of Science Abstracting and Indexing Services. *A Guide to The World's Abstracting and Indexing Services in Science and Technology.* Washington: National Federation of Science Abstracting and Indexing Services Report No. 102, 1963.

40. Preston, Lee E. and Norman R. Collins. "The Analysis of Market Efficiency." *Journal of Marketing Research*, May 1966, pp. 154-162.

41. ———. *Consumer Goods Marketing in A Developing Economy: The Case of Greece.* Athens: Center for Planning and Economic Research, 1968.

42. Price, Derek. *Little Science, Big Science.* New York: Columbia University Press, 1963.

43. "Research Studies in Patterns of Scientific Communication: The Roll of the National Meeting in Scientific and Technical Communication." *Information Storage and Retrieval*, August 1972, pp. 159-169.

44. Rosenberg, V. "Factors Affecting the Preferences of Industrial Personnel for Information Gathering Methods." *Information Storage and Retrieval* 3 (1967):119-127.

45. Scott, E. J.; H. M. Townley; and B. T. Stern. "A Technique for the Evaluation of a Commercial Information Service and Some Preliminary Results from the DRUGDOC Service of the Excerpta Medica Foundation." *Information Storage and Retrieval* 7 (1971):149-165.

46. Scherer, F. M. *Industrial Market Structure and Economic Performance.* Chicago: Rand McNally, 1971.

47. Sosnick, Stephen H. "Operational Criteria for Evaluating Market Performance." In Paul L. Farris, ed., *Market Structure Research*, pp. 81-124. Ames: Iowa State University Press, 1964.

48. Stern, Louis W., ed. *Distribution Channels: Behavioral Dimensions.* Boston: Houghton-Mifflin, 1969.

49. ——— and Adel El-Ansary. *Marketing Channels.* Englewood Cliffs: Prentice-Hall, 1976.

50. Stigler, George J. "The Division of Labor is Limited by the Extent of the Market." *Journal of Political Economy*, June 1951, pp. 189-190.

51. Utterback, J. M. "Innovation in Industry and the Diffusion of Technology." *Science* 183 (1974):620-626.

52. Weinberg, Charles. "The University Library: Analysis and Proposals." Stanford University, Graduate School of Business, Research Paper No. 169, July, 1973.

53. Wolff, Manfred E. "Primary Transmission of Scientific Information —Today and Tomorrow." *Journal of Chemical Documentation* 11:3 (1972):137-138.

54. Yokote, Gail and Robert A. Utterback. "Time Lapses in Information Dissemination: Research Laboratory to Physicians Office." The *Medical Library Association*, July 1974, pp. 251-257.

55. Zif, Jehiel and Dov Izraeli. "A Central Marketing Authority: The Case of Israeli Marketing Boards." In Charles C. Slater, ed., *Macro-Marketing Distributive Processes From A Societal Perspective*, pp. 238-247. Boulder : University of Colorado, 1977.

9
Marketing of Information Services: Strategic Considerations in Channel Selection

Alok K. Chakrabarti

The objective of this paper is to comment on some of the studies recently completed on the marketing of scientific and technical information (STI) and discuss some of the implications of the findings in developing marketing channel strategy. It should be pointed out at this stage that some of these studies were exploratory in nature and therefore many of the findings should be treated as tentative. The hypotheses and propositions put forward here need further testing.

Several types of organizations, private and public, both profit making and nonprofit, are currently involved in the storage, retrieval, dissemination, and distribution of STI. The Kennedy subcommittee (1975) in the U.S. Senate estimated that the federal government's involvement in the STI area amounted to about $1 billion per year. The importance of efficient and effective management of the STI systems to the success of technological activities related to energy, health care, defense, environment, and transportation cannot be overemphasized.

Scientific and technical information is transmitted from the originator to the intended user by various methods, the simplest and oldest one being a direct interpersonal channel. Interpersonal channels of communication between individuals can exist in both formal and informal contexts. Beyond the interpersonal channel, information is transmitted through recorded means; the oldest method is the printed format—books, journals, monographs are familiar examples. Recent technological advances in microfilming, electronics, and computers have made it feasible to record, retrieve, and deliver information in various formats. Application of these new technologies has revolutionaized the STI area and is shaping the information industry for the near future. King et al. (1976) estimated that the abstracting and indexing costs amount to about one-third of the total costs of organization and control of scientific and technical literature in the United States in 1976.

In discussing the STI industry one would quickly identify the inherent circularity between the originator and the user of STI in classical marketing sense. The originator of STI—the scientist/technologist—is at the same time also a user of STI. However, many other organizations participate to provide different functions such as printing and publication, storage, and retrieval, between the stages of the generation of scientific and technical information and its use.

Comer and Chakrabarti (1976) have proposed a multilevel model for STI distribution on the functional basis. This model, shown as Figure 9.1, consists of four levels. In the first level are the producers—the initiators of scientific and technical information. Initiators are scientists and engineers in academic and non-academic institutions. The second level in this model consists of publishers whose primary function is publishing STI. Examples of this level are publishers of journals, monographs, books and reports. The third level consists of various kinds of organizations whose primary functions are storage and retrieval of information. In recent years there has been a large growth of the services provided by organizations in this level. The last level consists of end-users of STI.

The linkage between the publishers and the users was formerly direct. Libraries performed the storage function and the librarian's job was largely custodial in nature. The interface between the user of STI and the librarian involved retrieval function. The classical cataloguing methods helped the retrieval process.

The application of computer technology has created a host of abstracting and indexing (A&I) services. These A&I services, sometimes known as secondary generators, depend on the primary publishers for the STI. Data bases have been generated by many organizations, both public and private. As depicted in Figure 9.1, STI from A&I services can be disseminated to the user

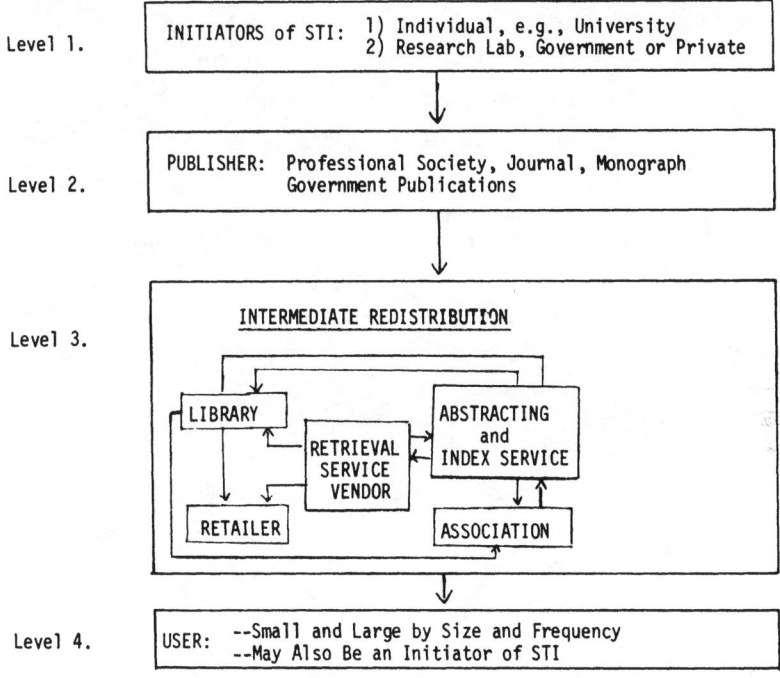

Level 1.

INITIATORS of STI: 1) Individual, e.g., University
2) Research Lab, Government or Private

Level 2.

PUBLISHER: Professional Society, Journal, Monograph
Government Publications

Level 3.

INTERMEDIATE REDISTRIBUTION

LIBRARY

RETRIEVAL
SERVICE
VENDOR

ABSTRACTING
and
INDEX SERVICE

RETAILER

ASSOCIATION

Level 4.

USER: --Small and Large by Size and Frequency
--May Also Be an Initiator of STI

FIGURE 9.1: CHANNELS OF DISTRIBUTION: STI*

* Reference: Comer and Chakrabarti (1976)

through many channels. Examples of the A&I services are Engineering Index, Chemical Abstracts and BIOSIS (Comer and Chakrabarti, 1976). An important A&I service is the one offered by the National Technical Information Service, which provides access to unclassified information generated through federally funded projects.

The "retrieval service vendors," shown at Level 3 in Figure 9.1, provide access to the various A&I services. For example, Lockheed Corporation has developed the DIALOG[TM] system through which a potential user can have access to any one of about 18 data bases in various fields. Systems Development Corporation has also developed a system that provides access to data bases. Subscribers to the services of the "retrieval service vendors" (e.g. Lockheed) include both direct user organizations, such as industrial organizations, and retailers who in turn sell to the direct users. There are a number of "retailers"—both non-profit and profit-making organizations—who cater to the needs of the users. Examples of the nonprofit organizations are the Knowledge Availability Systems Center at the University of Pittsburgh and the Computer Search Center at the Illinois Institute of Technology Research Institute in Chicago. In some instances, libraries have also been involved in acting as retailers of information. Northwestern University Library and several public libraries in the San Fransisco Bay area, are acting as retailers of A&I services for the customers.

This description briefly illustrates the different channels of distribution that are available for providing the A&I services. The natural question arises about the relative effectiveness of these various channels. The answer to such a question should be made in light of what we know from the literature in marketing. (For a list of references, the reader is referred to Comer and Chakrabarti, 1976, pp. 167-169.) The effectiveness of the different channels will depend on the nature of the market segment. In the STI field, there have been some studies recently concluded to define and explore the characteristics of different market segments. These findings will be useful in commenting on the choice of appropriate channels.

The BIOSIS study (Elias, 1976) identified five different groups of audiences for STI:

1. *Policy Manager*: R&D vice presidents and other top management personnel in technical areas. They do not use STI in their daily activities and therefore do not value STI unless a crisis occurs.
2. *Technical Managers*: R&D supervisors who are in direct contact with technical personnel and technical tasks and consequently are in need of STI. They try to keep up with the STI, but their involvement may vary depending on the nature of the organization.
3. *Practitioners*: This category includes a vast number of scientists and engineers who are involved in technical tasks. They are the actual users of STI. They are quite interested in the *form* in which the information is delivered to them.

A large fraction of the practitioner group, however, is involved in fairly routine technical tasks and therefore may not need sophisticated STI input.

4. *Intermediaries*: This audience is composed of people who deal with STI as their basic livelihood. Academic librarians and special librarians in various fields belong to this category. Many of the people perceive their role as being that of an unpaid salesperson for major data base suppliers. There are others who recognize that they have a personal stake in accepting STI as a valid and important research tool for both the industrial and academic research communities.

5. *Educators*: The college and university teachers belong to this category. Their information needs and requirements may be similar to that of practioners or technical managers, depending on whether they are actually involved in research projects or guiding research. As the BIOSIS study observed, their main preoccupation is with the theoretical aspects and the peer evaluation process.

If one pursues the observations made by the BIOSIS report, it becomes quite apparent that the different channels will have different levels of appeal as well as effectiveness in various market segments. The retrieval service vendors can cater directly to the intermediary, but for a policy manager it will be rather difficult to directly use the services provided by the retrieval service vendors. A commercial retailer, Editech Inc., reported that it is careful in selecting only top executives to make its sales presentations. (Comer and Chakrabarti, 1976, p. 63) It does not try to market its services to the corporate librarians, because librarians may perceive the services as competing with their own.

Wind and Grashof (1976) have developed different market segments based on the behavior characteristics such as price sensitivity, speed of information, concern for output format, extent of coverage, and nature of output. Applying sophisticated marketing research techniques, they identified five segments. They further investigated the buying process for STI services. As expected (based on the BIOSIS study [Elias,

1976]), they found that the buying center for the STI services is a complex organizational entity, the most important member being the R&D manager. Another important finding by Wind and Grashof (1976) is that

> the role of the librarian or information specialist has been over-rated. Generally speaking, the results showed that operating level information specialists have little input in the STI purchase decisions process. This result is particularly important since conventional wisdom in the STI field is that the information specialist is the primary agent for the purchase of STI.

Wind and Grashof were quick to point out that the information specialists played an important role in the use of the services and consequently are important contacts for the marketing of such services.

Another important finding relates to the different preferences for dealing with intermediaries. Small companies expressed the preference for dealing with consulting firms rather than the information service vendors directly. Large companies, on the other hand, had more structured STI organizations and preferred dealing directly with the vendors.

These studies point out the need for further development of understanding of the role of intermediaries in the distribution of STI services. The different types of channel members provide different types of services for different types of customers. The potential conflict between the various channel members remains a distinct possibility, as there is a common practice of selling between competitors. The conflict has not become apparent yet because the industry is still in its infancy.

Conclusions

The selection of appropriate channel strategy should be based on the careful consideration of the customer characteristics and their needs. As discussed previously, customers vary in terms of training and education with respect to STI services, preference for different formats, frequency of use, speed of document delivery, language used, etc. The different channels and intermediaries involved in the distribution of STI services vary greatly in terms of these dimensions. For example, a sophisticated, trained information specialist utilizes the large-scale service vendors effectively. A large R&D organization may be able to maintain such an information specialist on the payroll and subscribe directly to the large-scale vendors. However, firms of modest size and limited resources may find it useful to utilize the services of other small retailers. Some of the retailers perform the function of "repackaging" the information to suit the specific needs of their customers. This eliminates a great deal of redundancy in the process for the benefit of the customer. Time is also a very important element determining the economic value of information. The small retailers may provide personalized services in terms of information retrieval and document delivery.

As it is known in other areas of marketing, the market for STI services is not a uniform, homogeneous one. The market is heterogeneous and can be segmented in terms of some key characteristics of users and their behavior. Such a segmentation will help in the choice of appropriate channels for distribution of different STI services to different types of customers.

References

Comer, James M. and Alok K. Chakrabarti. *Channel of Distribution Strategy in the Marketing of Two Information Systems: A Comparative Study*. Report to the Division of Science Information. National Science Foundation grant No. SIS-75-13195, October 1976.

Elias, Arthur W. *Media Selection for Information User Training*. National Science Foundation Grant No. D.S. 175-19613-A02. BioScience Information Service, Philadelphia, PA; 1976.

King, D. W.; D. D. McDonald; N. K. Roderer; and B. L. Wood. *Statistical Indicators of Scientific and Technical Communication—1960-1980*. Figure 68. Rockville, MD: King Research Inc. Center of Quantitative Sciences, 1976.

Special Subcommittee on the National Science Foundation, Committee on Labor and Public Welfare. "Federal Management of Scientific and Technical Information Activities: The Role of the National Science

Foundation." Report. U.S. Senate, July 1975.

Wind, Yoram and John F. Grashof. *An Operational Experiment for the Marketing of Scientific and Technical Information Innovations.* National Science Foundation grant No. GN-42-271 and SIS-74-08626-A01, May 1976.

Part 5

Developing a Better Understanding of STI Marketing

Introduction

This summary section includes both papers that report the results of empirical studies and those that speculate on research questions and issues requiring further investigation.

John F. Grashof's paper reports on an empirical study in the STI domain. This study addresses some basic questions that must be answered before marketing theory can be effectively applied in STI. It also demonstrates the value of marketing research techniques in this environment and offers several prescriptive guidelines derived from the research. An especially important finding concerns the existence of an STI "buying center" which suggests the need to redefine the appropriate acquisition and user group within organizations.

The paper by William R. King and Jaime I. Rodriguez addresses a relatively neglected and difficult issue: the evaluation of information systems. Although not focusing on a conventional STI system, the system development and evaluation issues discussed in this paper apply to a wide variety of STI systems. The basic methodology offers a more sophisticated effectiveness- (in contrast to efficiency) oriented approach for evaluating scientific and technical information systems. This paper reports on an empirical study that implements a number of behavioral considerations long called for by others (see, for example, Kochen, 1976; Samuelson, 1974; and Allen, 1977).

In his paper, Franco Nicosia expresses an overall point of view concerning the potential for the application of marketing theory to STI, and poses some research issues to be resolved. This point of view is very provocative in its challenge of commonly held assumptions. Moreover, the consequences of implementing the implications of Nicosia's view would be substantial, at least in the sense that many people would have to change much of their current thinking about the transfer of scientific and technical information. The cogency of this view makes it very difficult to ignore and the potential benefits warrant careful consideration.

Joel Goldhar sums up the conference and the volume in his presentation of a research agenda. Clearly the closing note is a positive one, but Dr. Goldhar makes it clear that the proposed marriage between marketing theory and STI will not be easy to achieve.

References

Allen, Thomas J. *Managing the Flow of Technology: Technology Transfer and the Dissemination of Technological Information Within the R&D Organization.* Cambridge, MA: MIT Press, 1977.

Kochen, Manfred. "Can the Behavioral Sciences Contribute to the Foundations of Information Sciences?" *Proceedings of the American Society for Information Science,* 1976.

Samuelson, Kjell. "Information Models and Theories—A Synthesizing Approach." In Anthony Debons, ed., *Information Science: Search for Identity,* pp. 47-67. New York: Marcel Dekker, Inc., 1974.

W.R.K.
G.Z.

10

An Experiment in the Application of Marketing Theory to the Marketing of STI Products and Services

John F. Grashof

Background

The domain of scientific and technical information (STI), in its broadest sense, includes all aspects of the generation, storage, retrieval, and dissemination of data, facts, and information of a scientific or technical nature or about scientific or technical matters. The STI industry is somewhat more narrowly defined. For example, proprietary information is generally excluded, although the information activity within a firm will often be responsible for managing that firm's proprietary information. STI is made available to others through the entire range of informal and formal communication mechanisms. Scientists working in similar areas talk with each other on the telephone and at meetings, write letters to each other, and send one another copies of working papers and intermediate reports. The "invisible college" network is fast, responsive, and meets some of the information needs of its members.

It is, however, with the formal system that the STI industry, at least the firms in the information business, is most concerned. Information is made public through articles in thousands of journals and reports and in papers delivered at hundreds of conferences. Some of this information will be needed by someone, unspecified, at some unspecified time. The person needing the information may not know where to find it, where it originated, or even that it exists. Only some of the people needing the information will have the desire, the knowledge, and the ability to seek out this information.[1]

The business of the STI industry is making available the

information needed by scientists and engineers. To accomplish this a complex delivery system exists beginning with the means of publication and extending through a variety of intermediaries to the user. Dr. Etgar's paper in this volume discusses the various channels in detail; Figure 10.1 presents a summary diagram of STI delivery patterns.[2]

The National Science Foundation, NASA, private foundations, professional societies, and trade associations have been extensively involved in developing and supporting STI products and services. It is through funding from NSF and others that Chemical Abstracts and Engineering Index were begun. NSF funding is also, in part, responsible for the development of machine-readable formats of these sources and the computer software to search for and retrieve information from them. With the refinement of computer searching capability, the STI industry had the ability to provide fast, accurate bibliographic information to anyone who had a computer terminal.

The information needs of the STI user community are often grouped in four categories: (1) retrospective searches, (2) current awareness, (3) browsing, and (4) items of specific information or data. A fifth category, the solution to a complex problem, is important but not generally included because the use of consultants or internal experts is usually required.

Information products and services have come into the market to meet the needs in all of the above categories. The services range from Lockheed's DIALOG, a computer-oriented bibliographic search service, to *Current Contents*, a booklet containing photocopies of the tables of contents of a set of current journals. The products offered vary from general STI to specific areas such as ICI's patent search service and the National Library of Medicine's MEDLINE.

The greatest adoption of advanced STI products, services, and technology has been by large, high technology corporations.[3] Not surprisingly, chemical firms and pharmaceutical producers are among the highest users, often having very sophisticated in-house information systems. Indeed, these firms are often directly involved in the development of improvements in information products and services.

The potential benefits from the STI products and services

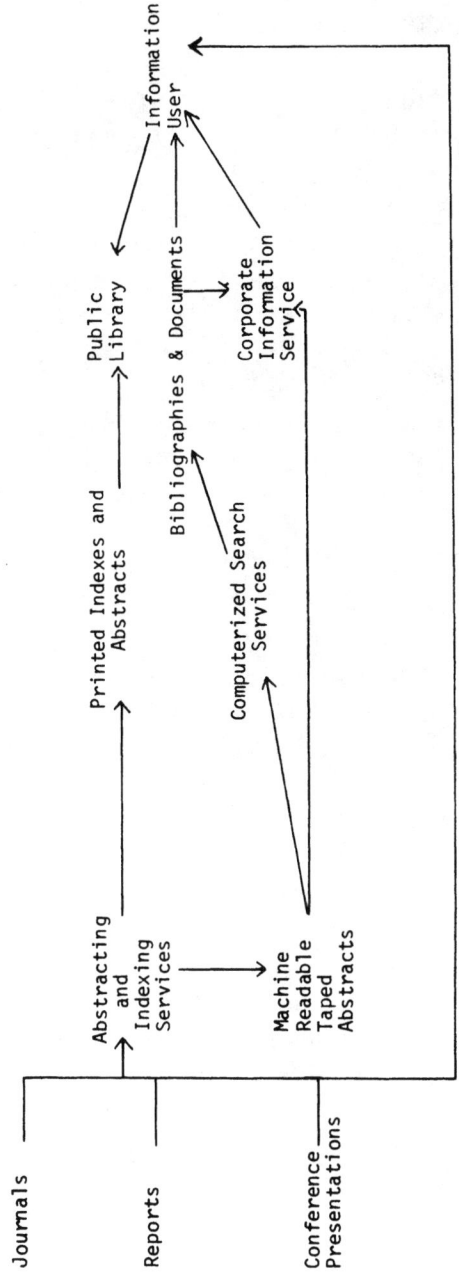

FIGURE 10.1: A SUMMARY OF STI DELIVERY PATTERNS

available have not yet been reached. Inexpensive, readily available technical information should also be of significant value to small and medium-sized manufacturing concerns. However, the adoption of information products by these firms has been very low. The reasons are many but include lack of knowledge of what is available, reluctance to spend money for information, and the need for more than bibliographic references.

The lack of adoption of STI products and services has a secondary impact on the industry: that of making it extremely difficult for information firms to survive. Most firms began with either direct or indirect subsidy from federal or trade association sources. Because volume has been low, prices had tended to remain rather high. This, in turn, discourages the adoption of products by small and medium-sized firms.

Problem Definition and Research Questions

The general problem to which this research was addressed was how to encourage greater adoption (or market penetration) of STI products and services by small and medium-sized firms. This problem was approached as a marketing problem and the research was to apply modern marketing theory and technology to the STI industry to try to increase penetration in the small and medium-sized manufacturing firms.

To guide the research a series of questions was posed:

1. What is the nature of the market for STI products and services?
2. What are the market segments for STI as defined by the attributes of the products and services they need?
3. What is the buying process, including the roles and responsibilities of the members of the buying center?
4. What are the potential supply systems and marketing techniques that can be used to satisfy the small and medium-sized manufacturing firms?

Based on these research questions a two-year study was funded · by the User Requirements Program of the Division of Science Information of the National Science Foundation.

Methodology

The methodology suggested by the study involved two types of activities. The first activity, or phase, was a comprehensive market research project to study the market for and the marketing of STI. The second phase, Year 2 of the study, was to attempt to implement in a realistic operating environment the findings of the first year's market research.

A significant amount of research has been conducted in the STI industry, primarily by persons in the information field.[4] Some studies of information users have been conducted but few, if any, have examined the buying process.[5] Studies of differences among users and market segments have tended to focus on mechanical considerations such as system protocol (computer commands) or use behavior rather than on the nature of information needs or the utility of various information product attributes.[6] Similarly there have been some studies of the marketing of information but these have tended to focus on census type questions such as, "How many information services are there?"[7] Few, if any, studies have focused on the marketing behavior of information services, or the channels of distribution used.

Because of the lack of market and marketing studies—studies conducted by marketing professionals using current theory and technology—the first phase of the project had the objective of filling in the gap in basic knowledge. The project designed included STI service marketing activity, STI buyer behavior, and a market segmentation study.

Once the market research was completed, the field or operational experiment (Year 2) could be conducted. The design called for establishing an operating information service that would actually develop information products and market them to real customers. All aspects of an STI provider were to be subject to experimental control including products, promotional activities, and means of distribution. In this way the performance of the information service using alternative marketing mixes could be evaluated.

The Year 1 Study

The preliminary (first year) investigation included personal and telephone interviews with industry leaders, focus group interviews, and on-site visits to STI information centers in major corporations—all aimed at providing the background essential to the design of the marketing and segmentation study.

Based on the results of this preliminary investigation, the major study was designed. The general objectives were to study the buying practices of STI users and potential users, the sources of products and product information, and potential market segments.

Sample Design

A stratified sample of 171 small, medium, and large firms was selected from the nearly 17,000 manufacturing concerns in the Commonwealth of Pennsylvania. The firms were selected to include representation from each of the 11 SIC codes identified by the National Science Foundation as "R&D intensive." The sample characteristics are presented in Table 10.1.

Questionnaire Design

The final research design called for personal interviews with representatives of the firms included in the sample. To guide the interviewers a 20-page questionnaire was designed. The questionnaire included five sections:

1. Investigation of the buying center
2. Data on the nature and extent of STI marketing activities
3. Respondent's awareness and use of STI
4. Conjoint measurement task
5. Demographic data on the respondent's firms

For the investigation of the STI buying center, each respondent was presented with a matrix composed of 11 aspects of a buying decision by 9 managerial/technical positions within a firm. The interviewer asked each respondent to rank the importance of each managerial-technical position in each stage of the buying process. The ranking task was performed twice by each respondent—once for an internally generated information need and once as a response to a visit by an STI product salesperson.

The data on STI marketing activities and the respondents'

TABLE 10.1

NUMBER OF FIRMS BY SIZE AND NUMBER OF RESPONDENTS BY POSITION IN FIRM AND SIZE OF FIRM

RESPONDENTS' POSITIONS

SIZE	President	Controller	Plant Manager	Quality Control Manager	Purchasing Agent	R & D Manager	R & D Scientist	Research Librarian	Director of Information	OTHER	TOTAL
SMALL (0 to 99 Employees) n = 60	13	4	17	8	7	24	6	5	2	8	94
MEDIUM (100 to 999 Employees) n = 57	1	5	8	6	1	38	20	12	3	11	105
LARGE (1000 or more Employees) n = 54	2	0	2	6	1	25	7	17	8	7	75
TOTAL	16	9	27	20	9	87	33	34	13	21	274

use of STI products were elicited by straightforward questions. The respondents were asked to rank the frequency and usefulness of several alternative sources of information on STI products. One section also asked them to evaluate alternative information suppliers. The questions on STI use and awareness were aimed at both the respondents' personal experiences and at the extent and sophistication of their firms' STI activities.

The research approach used to segment the market on the basis of the importance of and sensitivity to changes in product attributes was conjoint measurement. This approach requires respondents to rank order product concepts.[8] The rank orders are then analyzed and the market segments are identified. For this task, a set of 12 attributes of STI products and services was developed based on the preliminary research and experts in the STI field. Each attribute had from 2 to 9 alternative levels.

Using an orthogonal array approach (a type of experimental design that leads to a minimum of choice objects for a multi-level multiple-attribute product concept) a fractional factorial design was developed.[9] This design resulted in 20 combinations of factors and levels. These 20 combinations were printed on cards and the respondent was asked to order these hypothetical STI products from most preferred to least preferred.

The final section of the questionnaire asked for demographic data on the firms. Since some of the respondents were scientists and engineers and others were information specialists or managers two different versions of the demographic section were used.

Data Collection

Following the pretest and revision of the instrument the data were collected. Professional interviewers conducted the interviews, each of which lasted about one and one-half hours. The data were collected, the questionnaires edited, and the data was keypunched.

Data Analysis

The data were then analyzed in total, by type of respondent, and by the size of the respondent's firm.

A variety of types of crosstabulation analysis was conducted based on the research question asked and the nature of the data collected. A Thurstone Case V program was used to scale the buying center data.[10] Frequency tables and discriminant analysis were used on the marketing data and the usage and awareness data. A MONANOVA program was used to analyze the conjoint measurement data and a clustering program was used to identify market segments.[11]

Results and Conclusions

The following paragraphs and exhibits summarize the results of these analyses. More detailed discussions and interpretations of the analyses are presented elsewhere (as noted).

The STI Buying Center[12]

The data in Table 10.2 represent the Thurstone Scale values obtained from all respondents for situations A (an outside supplier offering a product) and B (an internally derived STI need). The data represent the relative importance of each position on each aspect with the highest scale value for each aspect underlined. The results of this analysis validated the hypothesis of multiple buying influence. The single most influential position is the R&D manager who received the highest scale value on 8 of the 11 aspects. However, there are several aspects where others were most important, such as "budget," where the controller was most influential and "negotiating with suppliers," where the purchasing agent had the greatest influence.

Most research on organizational buying suggests that there are differences in the roles and responsibilities of members of the buying center in different buying situations.[13] The results in Table 10.2 indicate that the same position has responsibility for the aspects of the buying process under both internally an l externally initiated consideration of the purchase of an STI product or service. Further, the relative scale values of all of the positions are not very different in the two situations.

The results of the analysis of the data by size of firm show that there are significant differences in the buying centers of different sized firms with respect to importance and influence in the STI purchase decision process. For small-sized firms, the

TABLE 10.2

THURSTONE SCALE VALUES FOR ALL RESPONDENTS IN DIFFERENT PURCHASING SITUATIONS

Situations / Aspects of Purchase Decision	President		Controller		Plant Manager		Quality Control Mgr		Purchasing Agent		R & D Manager		R & D Scientist		Research Librarian		Director of Information	
	A	B	A	B	A	B	A	B	A	B	A	B	A	B	A	B	A	B
Recognize Need	a.	.145		0.00		.376		.189		.019		.644		.566		.076		.142
Search for Alternatives		.051		0.00		.249		.160		.106		.524		.482		.171		.183
Establish Contact	.051	-.019	0.00	0.00	.209	.140	.081	.112	.316	.241	.545	.472	.278	.281	.053	.110	.135	.141
Set Purchase Criteria	.031	-.029	0.00	0.00	.234	.224	-.031	.017	.258	.246	.416	.421	.181	.210	-.075	-.105	.031	.055
Evaluate Alternatives	.017	.001	0.00	0.00	.255	.209	.189	.129	.064	-.018	.569	.527	.394	.388	.051	.005	.154	.093
Determine Budget	-.223	-.301	0.00	0.00	-.216	-.257	-.502	-.543	-.306	-.340	-.161	-.178	-.422	-.446	-.514	-.546	-.402	-.437
Evaluate Suppliers	-.043	-.072	0.00	0.00	.130	.101	.056	.049	.298	.285	.361	.363	.250	.282	-.001	-.022	.085	.081
Negotiate with Suppliers	-.093	-.060	0.00	0.00	.036	-.006	-.070	-.042	.580	.566	.200	.200	.007	.089	-.150	-.070	.026	.047
Select Supplier	-.027	.003	0.00	0.00	.103	.113	.007	.001	.432	.512	.305	.321	.124	.170	-.060	-.034	.077	.065
Determine Usage	.023	.002	0.00	0.00	.311	.289	.150	.180	.042	.025	.553	.555	.534	.553	.051	.018	.115	.081
Post Purchase Evaluation	-.043	.002	0.00	0.00	.130	.221	.055	.091	.062	.097	.392	.407	.250	.339	-.014	-.025	.065	.052

a. The first two aspects were not appropriate for situation A where the initiative was from the outside supplier. The first step in that situation is to establish contact.

plant manager is the position having the highest scale value for the largest number of aspects. The R&D manager scale value is second. In medium-sized and large firms, it is the R&D manager who is considered to be most influential.

The same pattern of shifting responsibility for different aspects of the purchase decision process is also reflected when these data are analyzed by firm size. The breadth of the responsibility of the R&D manager appears to increase as firm size increases. This may reflect an increased budget responsibility of the R&D manager in the large firms.

The implications of this segment of the research are important for producers of STI products and services. To be effective, suppliers must be prepared to deal with a variety of positions within an organization rather than limiting their sales efforts to research librarians and/or directors of information. Suppliers must widen their range of contracts within the organizations of potential customers, and tailor sales approaches to the various positions involved in the purchase decision process.

It is also clear that suppliers need to modify their sales activities as a function of the size of the firm. In small firms, the supplier will often be dealing with the plant manager and/or the purchasing agent; for a larger firm the contact will probably be the R&D manager.

The necessity of dealing with a variety of different positions will probably require a modification of the sales presentation. It is unlikely, for example, that a presentation that would be effective for the director of information of a large firm would be as effective to the plant manager of a small firm. The plant manager has a broader range of problems to deal with and is likely to be less sophisticated concerning STI products and services.

STI Marketing Activity

The survey questions concerning the level of current STI marketing activity concentrated on the sources of products and product information. In one task the respondents were asked to evaluate several alternative intermediaries according to a variety of specific criteria. Table 10.3 shows the resulting rankings.

In Table 10.3 it is obvious that, except for cost, firms prefer to

TABLE 10.3

INTERMEDIARIES PREFERRED ACCORDING TO VARIOUS SELECTION CRITERIA

Alternative Intermediaries

Criteria	Public Library	University Library	Private Consulting Firm	Trade Association	Government Agency
QUALITY	5	3	1	2	4
COST	1	3	5	2	4
CONVENIENCE	4	3	1	2	5
SPEED	4	2	1	3	5
FLEXIBILITY	4	2	1	3	5

Rankings are 1 = most preferred to 5 = least preferred.

use private consulting organizations as their source of STI and trade associations are second. Government agencies, many of whom spend a great deal of time and effort in offering information, are the least preferred on most dimensions. The preference for consulting groups seems to reflect a need, particularly among smaller firms, for answers to questions, not simply bibliographic searches.

The data in Table 10.4 show overall preferences for suppliers by firm size. Despite the highest ranking on criteria, consulting firms ranked only second in overall preference, probably because of cost. The preferences by size show some differences with the greatest surprise being trade associations. One would expect small (size 3) firms to rank trade associations very high yet their percent is much less than large and medium-sized firms. This may be because trade associations provide competitive information of value to larger firms but not information to help smaller firms solve problems.

Another aspect of the marketing of STI investigated was the frequency and value of contacts with sources of information regarding available STI products and services. Table 10.5 summarizes the ratings of alternative sources as to their frequency and usefulness.

The most frequent means of contact was via journal articles with journal advertisements and direct mail also significant.

TABLE 10.4

OVERALL PREFERENCES FOR STI SUPPLIERS

Sources:	Total	Size 1	Size 2	Size 3
1. Public library	9.2%	6.9%	9.5%	12.2%
2. University library	20.4%	15.3%	24.8%	23.0%
3. Private consulting firms	29.6%	33.3%	23.8%	32.4%
4. Trade association	35.9%	48.6%	41.0%	27.0%
5. Government agency	4.9%	5.6%	3.8%	8.1%

Size 1 = small, Size 2 = medium, Size 3 = large

TABLE 10.5

RATINGS OF FREQUENCY AND VALUE OF PROMOTIONAL CONTACTS REGARDING STI SERVICES

Sources:	Total		Size 1		Size 2		Size 3	
	A*	B*	A	B	A	B	A	B
1. Journal Articles	40.5%	29.8%	39.4%	33.8%	42.9%	26.7%	41.9%	32.4%
2. Journal Ads	13.1	7.3	11.3	5.7	19.0	10.5	5.4	5.4
3. Direct Mail Ads	12.4	4.4	6.9	2.8	9.5	6.7	18.9	0
4. Trade Shows	2.6	2.9	2.8	4.2	1.9	5.8	2.7	1.4
5. Professional Meetings	5.8	13.8	8.5	8.5	5.7	14.3	4.1	21.6
6. Others in your firm	4.7	5.1	4.3	2.9	1.9	2.9	10.8	10.8
7. Your counterparts in other firms	1.8	5.5	2.9	5.7	1.9	5.7	1.4	5.4
8. Other people in your firm in positions similar to yours	3.3	8.0	1.4	5.8	2.9	11.4	5.4	6.8
9. Persons in other positions in your firm	2.6	2.9	4.3	5.8	2.9	1.0	0	2.7
10. Salesmen	8.8	14.9	15.5	18.6	7.6	15.2	4.1	8.1
11. Other (Specify)	4.4	5.5	28.6	55.6	42.9	14.1	25.0	31.3

* Column A represents frequency of contact and Column B represents the usefulness of the information source. The data represent percent of respondents ranking each choice as most frequent (or useful).

When evaluated as to usefulness, journal articles again proved to be most highly rated. Professional meetings were also highly rated, with a distinct influence of the size of firm on this result. In smaller firms only 8% rated professional meetings as the most valuable source while they were so rated by 14% of medium-sized firms and over 21% of the larger firms.

Other data showed that the firms are contacted frequently about STI availability. Over 66% of the respondents indicated a contact at least once a month with over 30% reporting more than one contact per week. Large firms were generally contacted more frequently than were small firms.

Use and Knowledge of STI Products

Since the current level of use and knowledge is an important input to the design of a marketing plan for STI services, a section of the survey questionnaire was devoted to these topics. Data were gathered on a number of measures of use, awareness, and the degree of sophistication of potential STI customers.

Table 10.6 summarizes the results of nine such measures. These data show the impact of size on use, awareness, and sophistication. On every measure, the degree of awareness and/or use increases as size increases. Further, the data indicate that medium-sized firms are more like large than small firms regarding STI awareness, use, and sophistication.

The Conjoint Measurement Study[14]

The conjoint measurement section was included to determine the relative importance of various STI product attributes and to identify potential trade-offs among attributes that might be used for product design. Figure 10.2 presents the utility functions for the entire sample.

An examination of these functions suggests that price is the most important determinant of the purchase of an STI system. The major disutility is associated with a very high price level. A change from the lowest level to the medium-low level results in a disutility of only 1.54. This could be compensated for by changes in a number of factors such as the period coverage or mode of distribution.

Attribute substitutability can be determined based on the

TABLE 10.6

STI AWARENESS, USAGE AND SOPHISTICATION BY FIRM SIZE

	Firm Size [a]			Total
	Small	Medium	Large	
n =	72	105	74	274
	percentage	percentage	percentage	percentage of total respondents
1. Firm subscribes to printed abstract	30%	49%	74%	52%
2. Firm maintains a file of internal research results	62%	82%	90%	79%
3. Subscribe to external SDI sheet	28%	38%	53%	39%
4. Produce an internal SDI sheet [b]	18%	33%	46%	33%
5. Seen an on-line search system demonstrated	25%	42%	50%	40%
6. Firm subscribes to an on-line system	0%	10%	13%	7%
7. Firm maintains an STI library	51%	64%	82%	67%
8. Firm has a written budget for STI	16%	45%	66%	42%
9. Have used an outside information service	26%	57%	66%	51%

a. Small = 0-99, medium = 100-999, large = 1,000 and over employees.

b. SDI sheet is selective dissemination of information -- a prescreened current awareness service.

utility functions. For instance, providing citation plus selected abstracts is less preferred than offering a citation plus an answer to a problem. The difference in utility between the two outputs is 1.71, which can easily be compensated for by offering a printed computer output instead of microfilm (associated with an increase of 1.78 in utility). Similar trade-offs can be evaluated for any combination of attributes.

Figure 10.3 provides a comparison among five utility segments that were developed by clustering the MONANOVA

FIGURE 10.2: UTILITY FUNCTIONS FOR TWELVE STI SYSTEM FUNCTIONS: TOTAL SAMPLE

Variables	Mean	Mean Utility Values 1 2 3 4 5 6 7 8	Relative Importance of Factor
Speed of Obtaining Information			
1　Few hours	6.43		
2　Within a working day	6.38		
3　2-3 days	5.39		15.1%
4　4-5 days	4.29		
5　More than a week	1.22		
Purchase Arrangement			
6　Lease data, outside help	0.90		
7　Lease data, outside help, internal	1.07		
8　Lease data, search internal	1.06		1.5%
9　Lease data, system internal	1.17		
10　Subscribe outside services	1.43		
Nature of Output			
11　List of citations	1.40		
12　Citation + selected abstracts	3.98		
13　Citation + selected documents	4.02		
14　Citation + selected abs. or docs.	4.42		
15　Citation + selected abs. & inters.	4.79		12.4%
16　Citation + selected abs., docs., inter.	4.82		
17　Citation + facts from documents	5.04		
18　Citation + answer to problem	5.69		
Output Format			
19　Photo copy	7.62		
20　Printed computer	6.40		
21　Microform	4.62		18.0%
22　Temporary display	3.36		
23　Verbal	1.32		
Mode of Search			
24　Manual	2.37		
25　Computer offline	3.09		2.1%
26　Computer on-line	3.01		

Figure 10.2 (Continued)

Variables	Mean	Mean Utility Values	Relative Importance of Factor
		1 2 3 4 5 6 7 8 9 10 11	

Distribution

27 Direct request receive direct	2.91		
28 Direct request receive thru service	2.23		
29 Direct request receive outside serv.	1.81		
30 Internal request receive direct	2.26		
31 Internal request receive co. serv.	2.14		6.6%
32 Internal request	1.48		
33 Outside request receive direct	1.73		
34 Outside request receive co. serv.	1.23		
35 Outside request receive outside serv.	0.63		

Mode of Payment

36 Fixed price per hour	1.66		
37 Price per inquiry	2.23		
38 Fixed price per item returned	2.19		2.0%
39 Subscription charge	1.82		
40 Flat rate per subscription	1.53		

Type of Supplier

41 Trade association	3.00		
42 University	2.54		
43 Public library	2.29		6.0%
44 Private/consulting	2.91		
45 Government	0.92		

Language

46 English	4.92		1.5%
47 Computer language	4.40		

Topical Coverage

48 Complete	5.01		1.9%
49 Narrow	4.35		

Period Coverage

50 Everything available	3.74		5.8%
51 Most recent three months	5.75		

Price

52 High cost	3.16		
53 Medium high cost	9.06		
54 Medium low cost	10.78		26.6%
55 Low cost	12.32		

results. An examination of these utilities suggests that there is no universally desirable STI system. Different firms do differ in their preference patterns for system characteristics. The marketing management decision concerns which of these segments management should select as the target market. The variables to be considered include the size and other characteristics of the segments that may affect the ease and effectiveness of reaching them, the possible product lines the supplier can offer, and the number of different STI systems one can expect to market.

Summary of Important Conclusions

A variety of useful information has resulted from the study. Three of the most important conclusions relate respectively to the buying center, the STI product, and product attributes.

The conventional wisdom in the STI industry held that STI was purchased by the firm's information specialist, with perhaps some input from scientists and managers. The results of the current research show that such is not the case. Firms purchase STI in much the same way that they purchase other goods or services through the operation of an organizational complex or "buying center." The buying center for STI includes as its primary member the R&D manager (except in small firms where it is the plant manager). Other managers having significant input to the STI purchase process are controllers and purchasing agents. Except in the very large firms the information specialist is little involved in the purchase of STI.

The second conclusion is that there is no single product (or service) that will meet all the needs of firms using STI. Some firms need extremely fast service while others have little or no such need.

The largest segment of the market, about 48%, is relatively insensitive to the price charged (within a reasonable range); other firms are quite price sensitive. Thus a supplier cannot develop one product, take it to the STI market, and expect to satisfy the whole market. A variety of products, or combinations of options, must be made available if an STI marketer wishes to serve a broad range of firms.

Another major conclusion is that there are some product attributes that are important to the whole market. The twelve

PERCENT OF RESPONDENTS	SEGMENT 1 48%	SEGMENT 2 8%	SEGMENT 3 20%	SEGMENT 4 11%	SEGMENT 5 13%
	OUTPUT FORMAT (15.9%)	OUTPUT FORMAT (23.9%)	PRICE (34.8%)	PRICE (40.2%)	SPEED OF INFORMATION (30.8%)
	NATURE OF OUTPUT (13.6%)	PRICE (20.7%)	OUTPUT FORMAT (21.6%)	NATURE OF OUTPUT (17.9%)	PRICE (26.6%)
	SPEED OF INFORMATION (12.8%)	NATURE OF OUTPUT (13.9%)	NATURE OF OUTPUT (10.7%)	SPEED OF INFORMATION (13.5%)	OUTPUT FORMAT (16.3%)
	PERIOD COVERAGE (10.9%)	SPEED OF INFORMATION (8.8%)	DISTRIBUTION (8.3%)	OUTPUT FORMAT (10.4%)	LANGUAGE USED (5.2%)
	TYPE OF SUPPLIES (10.1%)	PERIOD COVERAGE (7.1%)	SPEED OF INFORMATION (7.4%)	TYPE OF SUPPLIES (4.9%)	NATURE OF OUTPUT (4.7%)
	PRICE (8.9%)	TYPE OF SUPPLIES (6.5%)	TYPE OF SUPPLIES (4.8%)	DISTRIBUTION (4.1%)	TOPICAL COVERAGE (4.3%)
	DISTRIBUTION (8.3%)	DISTRIBUTION (5.8%)	PERIOD COVERAGE (4.6%)	TOPICAL COVERAGE (2.0%)	MODE OF PAYMENT (3.0%)
	LANGUAGE USED (7.2%)	MODE OF SEARCH (4.3%)	MODE OF SEARCH (3.4%)	PERIOD COVERAGE (1.8%)	MODE OF SEARCH (2.5%)
	TOPICAL COVERAGE (6.3%)	TOPICAL COVERAGE (3.9%)	MODE OF PAYMENT (1.6%)	PURCHASE ARRANGEMENT (1.7%)	TYPE OF SUPPLIES (2.4%)
	MODE OF PAYMENT (2.9%)	MODE OF PAYMENT (3.2%)	PURCHASE ARRANGEMENT (1.3%)	MODE OF PAYMENT (1.9%)	DISTRIBUTION (2.3%)
	PURCHASE ARRANGEMENT (1.9%)	PURCHASE ARRANGEMENT (1.9%)	LANGUAGE USED (1.0%)	MODE OF SEARCH (1.1%)	PERIOD COVERAGE (0.9%)
	MODE OF SEARCH (1.3%)	LANGUAGE USED (0.0%)	TOPICAL COVERAGE (0.7%)	LANGUAGE USED (0.5%)	PURCHASE ARRANGEMENT (0.9%)

FIGURE 10.3: RELATIVE IMPORTANCE OF THE TWELVE STI FACTORS BY THE FIVE UTILITY SEGMENTS

STI attribute sets that were tested are classified according to whether each is important to none or some of the firms tested. (See Table 10.7.)

TABLE 10.7.
Importance of STI Attributes

Important to None of the Firms	Important to Some of the Firms
Purchase Arrangement	Speed of Obtaining Information
Method of Distribution of Results	Nature of Output
Mode of Search	Output Format
Method of Payment	Price
Type of Supplier	Inquiry Language
	Topical Coverage
	Period Coverage

The Field Experiment: Year 2

Based on the results of the Year 1 market, buying center, and segmentation study, a field experiment was designed. For part of this current experiment, existing STI suppliers (metascience, CDC, and PENNTAP) are being used.

Six groups of approximately 1,000 firms each were selected from the manufacturing firms in Pennsylvania. CDC and Metascience were assigned to serve chemical and metals firms respectively, and the membership list was used for the trade association cell. The three other groups were selected at random from the Department of Commerce's list of manufacturers.

Four operating supply systems were established, as shown in Figure 10.4. Each supply system was assigned one set of 1,000 firms as a target customer group. The fifth group will receive promotional materials from all four suppliers to determine which supply system they would use if given a choice among the four. The sixth group is a control group and will be surveyed, one-half initially and one-half at the end, but will otherwise not be contacted.

A marketing and evaluation plan, as shown in Figure 10.5, will be used to guide the promotional activities of the suppliers and the measurements of the evaluation group.

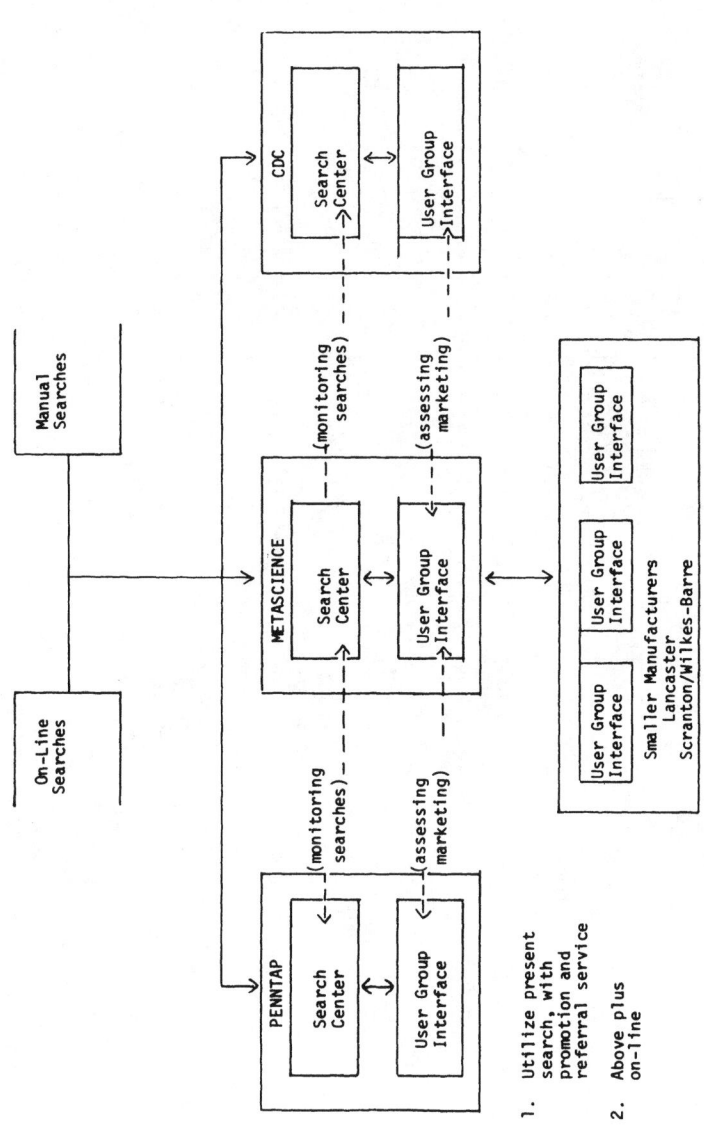

FIGURE 10.4: SERVICE AND SUPPLY SYSTEM

Channel Segment	Service	Market Approach	PERIOD					
			Users	Total	Users	Total	Users	Total
1	CBC n = 1,401	A - Direct mail only n = 500	1	2	1	2	1	2
		B - Direct mail with free search n = 400	1	2	1	2	1	2
		C - Direct mail with telephone follow-up n = 500	1	2	1	2	1	2
2	METASCIENCE n = 1,182	A - Direct mail only n = 400	1	2	1	2	1	2
		B - Direct mail with free search n = 400	1	2	1	2	1	2
		C - Direct mail with personal follow-up n = 382	1	2	1	2	1	2
3	PENNTAP n = 1,142	A - Direct mail only n = 642	1	2	1	2	1	2
		D - Special information article sent with direct mail piece n = 500	1	2	1	2	1	2
4	SMC n = 750?	A - Direct mail	1	2	1	2	1	2
		E - Demonstration to self selected subset at meeting of SMC	1	2	1	2	1	2
		F - Ad in SMC publication with "special" address for inquiries	11	2	1	2	1	2
5	All Four	A - Direct mail only	1	2	11	2	1	2
6	Control Group	E - No contact		3				4

1 = study of usefulness and satisfaction regarding information package used.
2 = study of attitudes and awareness regarding STI
3 = general attitude and knowledge survey
4 = post-experiment general attitude and knowledge survey.

FIGURE 10.5: MARKETING AND EVALUATION PLAN

Expected Results

The overall objective of the two-year study is to test whether marketing theory and modern techniques can have a positive impact on the penetration of STI products among small and medium-sized manufacturers. The Year 1 study showed that significant information can be gathered using marketing research techniques.

Year 2 is a test, under operating conditions, as to whether such information as gathered in Year 1 can be of value in improving the marketing structure and methods of STI suppliers. The study should also result in additional insights into the appropriate distribution channels and promotional mechanisms for STI products and services.

Conclusion

The study is significant because it marks an advance in the application of business techniques to the STI field. While some of the conclusions must be considered tentative until validated by other work, this change is an important step forward for the industry.

From a marketing theory standpoint, the study represents one of the few reported uses of operating firms in a research setting. In this project the objectives of the research were considered rather than the effect on profit. Therefore, research methods did not have to be modified to conform to good business practice.

Notes

1. Miles W. Martin and Russell L. Ackoff, "The Dissemination and Use of Recorded Scientific Information," *Management Science*, January 1963, pp. 322-336.

2. Ching-Chih Chen, "How Do Scientists Meet Their Information Needs?" *Special Libraries*, July 1974, pp. 272-280. See also H. Menzel, "Planned and Unplanned Scientific Communication," in *Proceedings of the International Conference on Scientific Information, Volume I* (Wash-

ington, D.C.: National Academy of Sciences, 1959).

3. Based on interviews with both suppliers and users of STI.

4. William D. Garvey, Nan Lin, and Carnot E. Nelson, "Communication in the Physical and the Social Sciences," *Science*, December 1970, pp. 1166-1173.

5. See, for example, D. E. Gushee, "Reading Behavior Chemists," *Journal of Chemical Documentation* 8 : 4 (November 1968):191-194.

6. H. J. Hall, "Technical Report on User Values in the Selection of Purchased Information Services," June 30, 1975, to August 31, 1976 (Linden, N.J.: Exxon Research and Engineering Company, Government Research Laboratories, Contract No. NSF C-1207 [1976]).

7. Manfred E. Wolff, "Primary Transmission of Scientific Information —Today and Tomorrow," *Journal of Chemical Documentation* 11:3 (1972):137-138.

8. For a more complete explanation of conjoint measurement, see Paul E. Green and Yoram Wind, "New Way to Measure Consumers' Judgement," *Harvard Business Review* 53 (July-August 1975):107-117.

9. P. E. Green, "On the Design of Choice Experiments Involving Multifactor Alternatives," *Journal of Consumer Research* 1 (Septembpr 1974): 61-68.

10. See J. T. Sims and T. Hammack, "TS-V: Thurstone's Law of Comparative Judgement (Case V)," *Journal of Marketing Research* 13 (May 1976):161-162.

11. J. B. Kruskal, "Analysis of Factorial Experiments by Estimating Monotone Transformations of the Data," *Journal of the Royal Statistical Society*, Series B, No. 27 (1965) pp. 251-263.

12. Based on J. F. Grashof and Yoram Wind, "The Boundaries of Buying Decision Centers," *Journal of Purchasing and Materials Management* (Summer 1978), pp. 23-29.

13. Yoram Wind, "Organizational Buying Center: A Research Agenda," in *Organizational Buying Behavior*, G. Zaltman and T. V. Bonoma (eds.) (Chicago: American Marketing Association, 1977).

14. Based, in part, on Y. Wind, J. F. Grashof, and J. D. Goldhar, "Market Based Guidelines for the Design of an Industrial Product," *Journal of Marketing*, July 1978.

15. As used here the term "utility function" refers to the relationship between the quantity of a specific attribute a particular product is perceived to possess and the utility or value of that product on a numerical scale. For example, consider the attribute weight in a household steam iron. As the weight of the iron increases the consumers utility evaluation of the iron increases (at least up to some point where it becomes cumbersome).

11
On the Evaluation of Scientific-Technical Information Systems

William R. King
Jaime I. Rodriguez

The concepts and techniques related to the evaluation of information systems are in their infancy. Sophisticated computerized information systems have not existed for a long enough period for anything else to be the case.

In the STI arena, most information systems are retrieval oriented. Much of the evaluation of these systems that has taken place has quite naturally been *efficiency oriented* in that it focuses on measures of accuracy, timeliness, redundancy, etc (for instance, see King and Bryant, 1971).

Ideally, STI system evaluation should also be *effectiveness oriented* to focus on such questions as:

1. Did the system produce what it was expected to produce?
2. Did the system have measurable impact on the behavior of the using organization or individual?
3. Did the system affect the values or attitudes of the using organization or individual?

Such questions are the essence of effectiveness assessment and until they are adequately addressed, critical issues of STI system evaluation will be unresolved.

This paper reports on an evaluation of an information system that was conducted in an experimental environment. The system being evaluated has many similarities to an STI system, although it deals with the organization's *competitive environment* rather than its *technological environment.* The system is retrieval oriented and it deals with data that are generated outside the organization, just as does an STI system. Therefore, the systems development and evaluation issues are analogous, if not identical, to the STI context.

Moreover, the study discussed here had the purpose of developing and demonstrating a more sophisticated evaluation methodology than that which is commonly used in information systems of any variety (see, for example, King and Rodriguez, 1978). This methodology, or some variant of it, is applicable to a wide variety of information systems—particularly those that draw on information from environmental sources.

The System Evaluation Methodology

Figure 11.1 represents a system evaluation methodology that shows assessments being made prior to the system design phase of the system life cycle, during the various phases of development, and subsequent to system implementation. These assessments are shown to be made in terms of attitude, value perceptions, information usage, and decision performance—all measures related to system effectiveness.

The simple system development process represented in the horizontal flow of Figure 11.1 is meant to be illustrative only. There are many more detailed specifications of the phases, or steps, to be followed in designing, developing, and implementing an information system (Murdick, 1970). The process shown in Figure 11.1 is meant only to show that assessments can be made prior to the commencement of the effort, at various intervals between the phases (however they may be prescribed), and subsequent to the implementation of the system.

Assessment in the Evaluation Process

The overall evaluation process in Figure 11.1 involves assessments that fall into four general categories: attitudes, value perceptions, information usage, and decision performance.

Attitudes and Value Perceptions. Attitudes and value perceptions are an important and often-neglected aspect of system evaluation. While most informal postimplementation assessments that are made of information systems are clearly attitudinal in nature, most systems are developed without formal

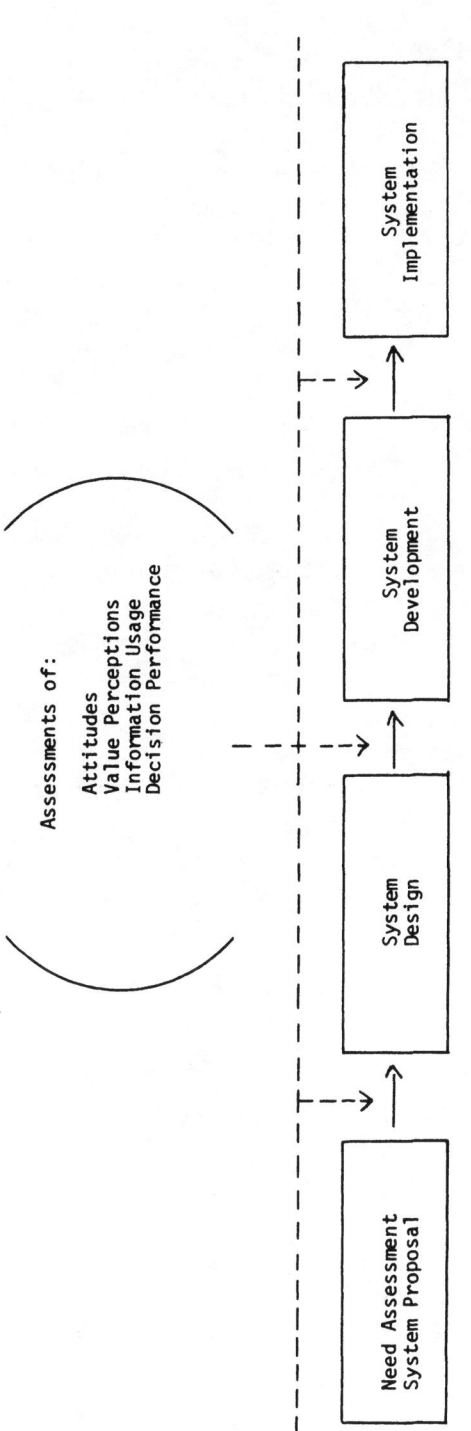

FIGURE 11.1: AN INFORMATION SYSTEM EVALUATION PROCESS

assessments of the attitudes of the users and the organization's managers.

Value perceptions are pragmatically distinguished from attitudes in the sense that they are more direct assessments related to the specific system. For instance, an answer to a question such as "How good is the system?" is a value perception in this sense, while an attitude is a more basic entity dealing with an individual's intrinsic beliefs and outlooks on the world (Triandis, 1971).

Decision Performance. Decision performance assessments reflect the quality of the decision-making process that may be supported by the system.

Informative Usage. A measure of system value that is related to decision performance is that of information usage. Even if a system does not impact on decision performance in measurable ways, it may reasonably be expected to affect user's information acquisition and usage behavior in identifiable ways. Since a user of a new system has, through his usage, altered his behavior, such an assessment must be made in more substantive terms. For instance, the assessment may be made in terms of whether the system has motivated the user to assess his choice situation more systematically, to make greater use of relevant information, or to use a decision model in a situation where he may previously have made an ad hoc choice.

The Evaluation Process

To evaluate a system, the four broad varieties of assessments may be made at the various stages of the design and development process and compared in two general ways:

1. Have attitudes, value perceptions, information usage behavior and decision performance of users changed from the presystem time period, through design and development, to the postsystem period?
2. Has system user behavior changed more than that of nonusers or have users benefited more than have nonusers?

Temporal Changes. Although all of the possible comparisons of the temporal variety undoubtedly would not be feasible in each system development situation, each such comparison should be considered as a possible element of a system evaluation.

In many cases, temporal assessments will be made only on a "pre-post" basis using measurements made before system design and after system implementation. However, if such assessments are to be useful in changing the nature of a system or the design-development process for a system under development, assessments made during the design-development process will also be necessary.

Predesign assessments of user attitudes can, in isolation, be useful to system designers. For instance, a predesign attitude assessment that shows clear negative user attitudes toward sophisticated systems might lead to a revised system concept, to greater inclusion of users into the design-development process, or to "sales" efforts that have the objective of achieving greater appreciation for the system by its potential users.

User Benefits. Evaluations of the second variety—those that assess the behavior changes or benefits accruing to system users relative to nonusers—are essential to ensuring that changes or benefits are not attributed to a system when they are really due to some other influence. For instance, a general positive shift in attitudes toward sophisticated computer systems might be caused by publicity concerning the effectiveness of new computers. If such an assessment were made for system users without comparison with nonusers, the shift might be attributed to the system when it is, in fact, an overall attitudinal change in the population—users and nonusers alike.

The System Evaluation Demonstration Study

The theoretical evaluation model of Figure 11.1 has been partially applied in a system development context. As with most practical applications of the conceptual framework of Figure 11.1, the measurements that have been made were selected on the basis of feasibility, cost, and relevance to the objectives of the system being evaluated.

The System Being Evaluated

A brief description of the system that served as the vehicle for demonstrating the evaluation model is necessary in order to fully understand the case study. The system is one that is designed to aid business managers in understanding and resolving competitor-related strategic issues. It is referred to as a "Strategic Issue Competitive Information System" (SICIS). A more detailed description of the system is given in Rodriguez and King [7].

The SICIS system is a variety of the "intelligent management information system" as described by King [3], or a mangement decision support system—in that it is designed to support the unstructured strategic and policy planning activities of an organization by providing the user with problem-related information that may go beyond that which he might otherwise be able to identify as critical to the making of a strategic choice. The emphasis of this system is the facilitation of competitive analysis and the identification of opportunities in the competitive environment by managers who are involved in strategic planning activities.

A key feature of the system is that SICIS is intended to be used by managers who are neither trained in competitive analysis nor in the use of sophisticated computer systems. If a system can be developed that is perceived to be useful and valuable by managers who are untrained in these areas, it can serve as a valuable tool in supporting a strategic planning process.

The SICIS system utilizes "strategic issues" representing strategic problem-related questions that the user can use to access competitive information in the SICIS data base. The strategic issues are modelled within the system so that each user query concerning an issue evokes an information structure model that defines, in hierarchical form, the information that is relevant to the query.

For instance, a manager using the system might make a query such as "What is the capability of Competitor X to introduce a new product in his Line Y next year?" and would be presented with system responses that successively identify subclasses of data determined to be critical to the resolution of this issue. For instance, such a query might produce a system response that would identify Competitor X's financial capability, production capability, marketing capability, and technological capability, as elements of the overall issue.

The user could then identify specific areas of interest, or he could indicate that he desires the "total picture" related to his initial question. A user indication of interest in the competitor's marketing capability would produce a system response indicating the availability of data on competitive distribution channel capacity, field sales capability, service capability, technical sales expertise, and a variety of other marketing-related areas. The system might also indicate its ability to provide projections of future market growth. Indications of interest in other areas would produce similar system responses which would, after each new user response, indicate successively more detailed sets of available data.

Thus, the SICIS system is much like an STI system or a retrieval system in terms of the information it makes available to the user. It allows the user to make queries and to refine them to obtain that which is needed and that which is decision relevant from the data base.

Figure 11.2 is a computer printout showing how the SICIS operates. In this instance, the system is programmed with data representing Tappan Corporation and a number of its competitors. These data are taken from a publicly available Tappan business policy case and a series of cases dealing with Tappan's competitors to avoid problems of confidentiality (Christensen, Andrews, and Bower, 1973).

The system user in Figure 11.2 has begun by identifying his interest in both Tappan and General Electric. From the strategic issue list, he has then selected Issue No. 2—"What resources limit the company's/competitor's ability to achieve goals?" The system responds by identifying six categories of constraints ranging from "financial conditions" to "imposed constraints."

The user may choose some or all of these categories for further elaboration. In effect, he can redefine the issue more narrowly than the one with which he initially inquired, or he can narrow it to encompass any, or all, of the six "subissues." In this case, the user selects the first and fourth subissues—those involving "financial conditions" and "marketing or distribution system" constraints.

Then, the system addresses each of these subissues in se-

ENTER CODES OF WHICH COMPETITORS TO EXAMINE
01—Tappan Co.
02—Design and Manufacturing
03—Sears Co.
04—General Electric Co.
99—All Competitors
1,4

ENTER ISSUE CODE OF DESIRED STRATEGIC ISSUE
IF NO MORE ISSUES DESIRED ENTER 00
2
    2—What resources limit the Company's/Competitor's ability to achieve goals?
        1—Financial conditions.
        2—Management constraints.
        3—Manufacturing efficiency.
        4—Marketing or distribution system.
        5—Research and development.
        6—Imposed restraints. .

ENTER CODES OF DESIRED COMPONENTS; IF ALL ENTER CODE 99
1,4
        1—Financial conditions.
          (1) Present loan levels.
          (2) Quantity of funds that can be financed internally.
          (3) Capital structure.

ENTER CODES OF DESIRED COMPONENTS; IF ALL ENTER CODE 99
2
        2—Quantity of funds that can be financed internally.

TAPPAN CO.
(millions)

	1970	1969	1968	1967
Sales	313	133·8	124·2	95·7
Profits	2·23	3·44	3·97	1·00
Return on Assets	2·8%	4·5%	5·3%	1·9%
% Profit/Sales	1·7%	2·6%	2·6%	1·0%
Ten Year ROA	5·9%	—	—	—
Ten Year Profit/Sale	—	2·8%	—	—

GENERAL ELECTRIC CO.
(millions)

	1970	1969	1968	1967
Sales	8,726	8,448	8,381	7,741
Profits	328	278	337	361
Return on Assets	00.0%	00·0%	00·0%	00·0%
Profit/Sales	3·8%	3·3%	4·3%	4·7%
Ten Year ROA	17·2%	—	—	—
Ten Year Profit/Sales	—	4·6%	—	—

D—Marketing or Distribution System
    (1) Own us. Independent distribution.
    (2) Aggressive or passive market organization.

ENTER CODES OF DESIRED COMPONENTS; IF ALL ENTER CODE 99
1
    (1) Own vs. Independent distribution.

TAPPAN CO.
Tappan counted with about 120 salesmen which sold to retailers, large builders, and independent distributors. Tappan was changing its distribution system and was setting up its own centralized and regional warehouses.
Tappan franchised its servicing capability to private companies.

GENERAL ELECTRIC CO.
    GE controls its own channels of distribution as well as service outlets and has a potential advantage in surrounding Sears with coverage using the GE Brand, Hotpoint and Penncrest Brand.

ENTER CODES OF WHICH COMPETITORS TO EXAMINE
01—Tappan Co.
02—Design and Manufacturing
03—Sears Co.
04—General Electric Co.
99—All Competitors

ENTER ISSUE CODE OF DESIRED STRATEGIC ISSUE
IF NO MORE ISSUES DESIRED ENTER 00

00
    THANK YOU FOR USING SICIS TRY US AGAIN

EXIT

quence. It first presents the next level of categories within the "financial constraints" subissue. These are identified as "present loan levels," "quantity of funds that can be generated internally," and "capital structure." Again, the user may select any or all of these. In this run, he chooses the second only and is presented with tables of financial data for both Tappan and GE.

Then the second subissue that he previously chose—"marketing or distribution system"—is presented to the user in terms of its two subcategories, "own versus independent distribution" and "aggressive or passive marketing organization." In this case, the user chooses only the former and is presented with narrative statements that depict this aspect of Tappan and GE's business activities. Since these particular statements are taken from dated publically available business policy cases, they are somewhat innocuous, but they do suggest the nature of the information that could be incorporated into the system.

After presenting these data, the system returns to its initial state and asks the user if he wishes to examine any additional firms or issues. In this illustration the user indicates that he has completed his use of the system.

The Evaluation Measures

The case study in question utilized specific assessments *in each of* the four general areas. *Assessments of attitudes and value perceptions* were made using an instrument developed by Schultz and Slevin [9]. This instrument entails both items that can be related to general attitudes and specific indications of the perceived value of the system. Table 11.1 describes both varieties of these measures.

Information usage behavior is assessed in the demonstration study in terms of the *amount of use* that is made of the system and the *substantive nature of the usage.* In the case of SICIS, the amount of use is measured by the *number of queries* that users make when they are faced with strategic issues on decisions for which the system can be of help. The substantive nature of system usage is assessed in terms of the correlation between the issues that are addressed during system usage and the inputs that were previously provided by users during the earlier phases of system design and development. Users were

TABLE 11.1

Schultz-Slevin Questionnaire Attitudinal Factors

  F1 - Performance: effect of system on manager's job performance
       and performance visibility.
  F2 - Interpersonal: interpersonal relations, communications, and
       increased interaction and consultation with others.
  F3 - Changes: changes will occur in organizational structure and
       people I deal with.
  F4 - Goals: goals will be more clear, more congruent to workers
       and more achievable.
  F5 - Support/Resistance: system has implementation support -
       adequate top management, technical and organizational support
       and does not have undue resistance.
  F6 - Client/Researcher: researcher understands management problems
       and works well with clients.
  F7 - Urgency: need for results, even with costs involved, importance
       to me, boss, top management.

Measures of Value Perceptions of System Users

  D1 - Probability that you will use the system.
  D2 - Probability that other managers will use the system.
  D3 - Probability that the system will be a success.
  D4 - Managers evaluation of the worth of the system.
  D5 - The level of accuracy you expect from the system.

asked early in the participative design-development process to specify which issues would be most usefully incorporated into the system. Their inquiries into these issues using the system were subsequently assessed and compared to their inputs. Such a matching provides an assessment of whether user inputs serve substantive purposes. An alternative hypothesis might be that user involvement in system design serves only psychological purposes in reducing anxiety and facilitating change.

Decision performance is assessed in the demonstration study in terms of user performance on a series of specific competitive-oriented strategic choice problems that were prepared and responded to by participating users. Their answers to these questions were evaluated by independent objective evaluators (professors of business policy) who "graded" both the substance and the thought processes that were used to justify user responses.

Using these measures, each of the four general varieties of assessment were made in the case study. Of course, as in any real system development effort, the situation did not permit all of the logically possible temporal comparisons to be made. Neither did it permit the decision performance assessments to be made in terms of a sequence of real-world choices made over a period of time. However, those assessments that were feasible serve as a basis for a system evaluation that goes well beyond those that are usually made.

The Evaluation Context

The system evaluation demonstration study that is reported on here was conducted for the SICIS system using 45 experienced manager-users in a simulated business environment. The managers were enrolled in a part-time Master of Business Administration program and had completed virtually all of the program requirements. The simulation was made a part of a "capstone" course, "Integrated Decision Making," which utilizes business policy cases as a primary teaching vehicle.

The class had spent six class hours plus preparation time on a series of cases from the appliance industry before they were involved in the SICIS exercise. One group of subjects participated in the system design and then used the system. A second group which did not participate in the design also used the system. A third (control) group did neither.

The subjects were all experienced managers who were assigned to the three groups on a randomized blocking basis (Cox, 1958). This assignment ensured that the groups were alike with respect to their proportional composition of managers with much, some, or little experience in strategic competitive analysis—the focal point for usage of the SICIS system.

The managers were not told that they were involved in an experimental evaluation of the system until the process was complete. Rather, the exercise was an integral part of the course that the managers perceived to require unusual scheduling in order for them to gain adequate access to the computer terminals on which they used the SICIS system. After the experiment was completed and all assessments were made, the control group, which had neither participated in the design nor used the system prior to that point, was permitted to use the system. This ensured that all groups were ultimately treated alike and that the educational objectives of the course were realized.

Evaluation Results

The results of the demonstration study evaluation may be expressed in terms of a series of "hypotheses" which were tested using the measures of attitudes, perceived value, information usage, and decision performance.

Attitudinal Hypothesis. The attitudinal hypothesis related to the positive change in attitudes that might be expected from system users. The change was measured on a pre-post basis. The "pre" measurement was taken after the managers had only a brief introduction to the objectives and nature of the SICIS system. The "post" measurement was made after they had an opportunity to make use of the system in responding to a set of business policy questions that were posed to them.

The attitudinal hypothesis, stated in null form, was that attitudes toward the system would not change favorably after use of the system in comparison to pre-use attitudes. Two groups of managers—one composed of those who had used the system and one of those who had not—were administered the Schultz-Slevin [9] instrument on a pre-post system usage basis. The control group's attitudes did not change significantly in terms of the "gain scores" (pre versus post measurements) on the seven Schultz-Slevin attitudinal factors. However, the "experimental" group—those who had used the system—had significant positive attitudinal change in the "performance" factor (F1) and the "client researcher" factor (F6) (see Table 11.1). The other five attitudinal factors did not show a statistically significant difference in gain scores. Table 11.2 shows the group means, standard deviations, and significant levels for the test in terms of each of the seven factors.

Perceived Value Hypothesis. A hypothesis of perceived value was addressed using the Schultz-Slevin [9] dependent variables in Table 11.1. Again, gain scores for users and nonusers were compared to test the hypothesis (in null form) that the value perceptions of system users would not change relative to the value perceptions of nonusers.

Table 11.3 shows the means, standard deviations, and significance levels that reflect significantly differences in the "use" and "worth" variables with no significance being attached to

TABLE 11.2

	User Average (Standard Deviation)	Non User (Standard Deviation)	Level of Significance
F1 (Performance)	.313 (.193)	.035 (.345)	.0189
F2 (Interpersonal)	.865 (.60)	1.15 (.556)	.3290
F3 (Changes)	.166 (.454)	-.023 (.886)	.4159
F4 (Goals)	.623 (.420)	.447 (.30)	.1159
F5 (Support/Resist)	.260 (.438)	.315 (.277)	.2902
F6 (Client/Researcher)	2.40 (.303)	2.21 (.551)	.0567
F7 (Urgency)	.433 (.312)	.448 (.305)	.3997

TABLE 11.3

	User Averages (Standard Deviation)	Non User Average (Standard Deviation)	Significance Level
D1 Use	.064 (.333)	-.07 (.300)	.073
D2 Others	.007 (.237)	-.01 (.246)	.363
D3 Success	.050 (.176)	-.014(.23)	.242
D4 Worth	.014 (.092)	-.07 (.170)	.109
D5 Accuracy	.050 (.105)	.107(.175)	.205

the differences in changes in the other three variables. This implies that users perceive the value of the system to be greater in terms of an overall assessment of its worth than do nonusers, but the same difference between users' and nonusers' perceptions does not hold in terms of the usage of other managers, system success, and "accuracy." (Note that the Schultz-Slevin instrument was initially developed for use in evaluating a forecasting model. The "accuracy" variable has much more relevance there than with SICIS. Because the authors were desirous of using an established and validated instrument, it was used despite the limited relevance of a few of its items and factors to SICIS.)

Information Usage Hypothesis. The hypothesis dealing with the amount of information usage involved a comparison of design participants and nonparticipants, rather than, as in the previous hypotheses, of system users versus nonusers. No pre-versus-post information usage evaualtion was performed.

One group of managers participated in system design by critically evaluating a list of strategic issues that the system might be designed to address, by suggesting other issues that might be incorporated into it, and by speculating on the relevant information structure models that might be built into the system for each issue. Another group had no such participation opportunity. Both groups used the system as an aid in responding to a series of business policy questions. The two groups were compared with respect to the amount of use (number of inquiries) they made of the system. This was measured using protocols that were generated by each user from the interactive computer system on which the system was implemented.

The two groups (participants and nonparticipants) showed no significant difference with respect to the amount of system usage in this study. Participants made an average of 7.73 inquiries while nonparticipants made an average of 7.29 inquiries—an insignificant difference.

This result might be viewed as a positive evaluation of the system since it implies that less-knowledgeable managers (nonparticipants) find the system to be as easy to use as do those managers who have participated in its development. Such a characteristic is a desirable one for a system of the SICIS variety, since it is intended for use by managers who are not trained in either computer science or competitive analysis.

A second hypothesis related to information usage concerns the degree to which the substantive input of managers, as provided during the design-development process, is related to the substance of actual system use. Presumably, a correlation between input and usage would reflect well on both the design process and the system.

As a part of the design process, participants were asked to rate each issue on a five-part Likert-type scale in terms of its relevance to the simulated business context. Subsequently, after the system had been developed and the design participants had used it, their protocols were analyzed to ascertain which issues had been inquired into by them in responding to the business policy questions.

The (null) hypothesis (that the inputs provided in the design process would not be reflected in system usage) was rejected using a Kendall Rank Correlation test, which assesses the similarity in rank between the values placed on the issues during the design process and the rank in terms of frequency of actual use. The Kendall coefficient of correlation, Tau, was 0.6277 giving a significance level of 0.0016. Thus, the null hypothesis was rejected and it was concluded that there is good reason to believe that the two rankings are highly related.

This finding serves to validate both a participative design process and the system that emanates from it. As such, it is an important criterion that can be used for evaluating SICIS.

Decision Performance Hypothesis. The "bottom line" assessment of a system—particularly one like SICIS which is directly related to strategic decision making—is improved decision performance.

System users were compared with nonusers in the demonstration study to determine if they were significantly different with respect to their performance on a number of business policy questions related to the simulated business. Three business policy professors evaluated their responses. The hypothesis test was based on the overall average across the three professor's grades because the three were shown to be consistent—e.g., an interjudge reliability test showed that the three professors did not differ significantly in their ratings.

The overall ability of the user and nonuser groups was also shown to be comparable in terms of their overall quality point average in a Master of Business Administration (MBA) program that most were completing. Thus, in terms of this measure, at least, the user and nonuser groups were similar.

The average score achieved by the user and nonuser groups was 21.04 (with a standard deviation of 5.28) and 21.64 (with a standard deviation of 4.12), respectively. This leads to the acceptance of the null hypothesis that system users and nonusers do not perform differently in addressing decision-related policy issues.

Thus, the "bottom line" basis for evaluating the SICIS system did not bear out its value in this demonstration. However, this may be attributable to the "one-shot" nature of the decision performance measurement. Such assessment is at odds with the typical purpose of a decision-support system—that of providing *continuing* support for a sequence of decisions. Hence, the partial application of evaluation methodology in this demonstration study may well be at fault in failing to adequately assess decision performance.

Summary

The demonstration study demonstrates the feasibility of assessing system effectiveness on the basis of a comprehensive conceptual model using measures of attitudes, value perceptions, information usage, and decision performance. Although the demonstration study was conducted in a simulated environment, similar measurements can as well be made as a part of real-world system design-development-implementation processes.

Just as is likely to be the case in most system-evaluation efforts, the SICIS system that was evaluated in this study was not evaluated as highly as might have been hoped for by its developers. However, evaluations such as these serve to guide further system evaluations efforts as well as to suggest system refinements and characteristics that future systems should entail.

The system evaluation process is thus a dynamic one in which each systematic evaluation effort can lead both to better information systems and to improved evaluation methodologies. Hopefully, this will lead to the application of such evaluation methodologies on a broader scale—in real-world user organizations as well as in other environmental information system contexts such as STI and the regulatory and social environments of business.

References

Christensen, C. R.; D. R. Andrews; and J. L. Bower. *Business Policy: Text and Cases* (3rd edition), pp. 323-431. Richard D. Irwin, Inc., 1973.

Cox, D. R. *Planning of Experiments.* John Wiley, 1958.

King, D. W. and E. C. Bryant. *The Evaluation of Information Systems.* Information Resources Press, 1971.

King, William R. "Intelligent Management Information Systems," *Business Horizons,* October 1973.

King, William R. and David I. Cleland. "Environmental Information Systems for Strategic Marketing Planning." *Journal of Marketing,* October 1974.

King, William R. and Jaime I. Rodriguez. "Evaluating Management Information Systems." *MIS Quarterly,* September 1978.

Murdick, R. G. "MIS Development Procedures." *Journal of Systems Management,* December 1970, pp. 22-26.

Murdick, R. G. and J. E. Ross. *Information Systems for Modern Management* (2nd ed.). Prentice Hall, 1975.

Rodriguez, J. I. and W. R. King, "Strategic Issue Competitive Information Systems." *Long Range Planning,* February 1978.

Schultz, Randall and Dennis Slevin, eds. *Implementing Operation Research —Management Science.* Chap. 7. New York: Elsevier-North Holland, 1975.

Siegal, Sidney. *Nonparametric Statistics.* McGraw-Hill, 1956.

Triandis, H. C. *Attitude and Attitude Change.* John Wiley, 1971.

12
On the Buying and Using of Scientific and Technical Information by Organizations

Franco Nicosia

One of the founders of operations research, C. West Churchman, once taught me that the simplest definition of an organization—an organized set of people and machines—is also the most difficult to describe. That all organizations consist of people and some minimum of human artifacts (machines) is true, but that such people and machines are organized is debatable. In fact, in our daily contacts with organizations we often conclude that many of them are disorganized. Moreover, according to some philosophers—and probably to a few students of consumer behavior—organizations have no organization at all, for they behave stochastically.

It is an understatement to say that I feel hesitant about sharing with you some thoughts I have on the subject. To begin with, if it were true that the Scientific and Technical Information (STI) industry—especially the senders and distributors of information—were making good profits and returns on investments, then researchers should ask the industry representatives how we can buy a share of their enterprise. On the basis of what I know about the marketing of books, journals, and technical magazines, and on the basis of what I experience as a seeker of scientific and technical information, however, I would conclude that there is room for substantial improvement not only in terms of profits and return on investment, but also in corporate planning and management decision making by the firms in the STI industry.

I want to express my gratitude for the suggestions by Dair Gillespie and Jon Minkoff.

Can organizational points of view help the STI industry? And can they help the formulation of public policy? At present, neither the viewpoint I shall discuss nor any others can help the practitioners. My primary purpose is to attempt to rearrange our way of thinking of an organization as a buyer and user of STI so as to understand the roots of the industry's practices and problems. My secondary purpose is to elaborate some parts of the framework presented earlier in this volume by King and Zaltman; to build upon the comments on individual user behavior by Sheth; and to provide some inputs to the discussion of information-acquisition strategy by Rothberg, new product design by Pessimier, distribution by Etgar, market segmentation by Wind and Thomas, and evaluation by King and Rodriguez. Hopefully, my comments may also provide the basis for novel insights—perhaps even quasi hypotheses—that may lead to research relevant to both the STI industry and public policy-makers.

There are two simple reasons why I cannot promise more. First, the basic disciplines have not yet delivered immediately useful insights. From the still-advanced work by March and Simon (1958), to the very innovative work on small groups in laboratory situations by Mackenzie (1976) and to the verbal and empirical work in modern contingency theories, basic questions about organizational structure and behavior are still being investigated.

Second, the marketing literature has, until recently, avoided the systematic study of the organizational qualities of group decision processes. On the buying side, for instance, only recently have any efforts been made to go beyond the dos and don'ts listed in the traditional industrial procurement literature (see, for example, Webster amd Wind [1972], Sheth [1976], and Nicosia and Wind [1977]). From our point of view, it is unfortunate that this literature seems to evolve into the study of materials management rather than the study of the organizational features of the buying process. On the selling or marketing side, the paucity of interest in the organizational properties of such activities is even more striking (see, e.g., the discussion in Nonaka [1972] and Nonaka and Nicosia [1977]).

These disclaimers should not discourage us. On the contrary, they present an opportunity and a challenge. While some points I make may state the obvious, the total picture should tell us that we must start at the beginning—namely, we must become aware that the obvious is not obvious at all and must thrive on the opportunity to develop new perspectives, new constructs, and new data.

The potential contributions of marketing theory to private and public policymakers depend on the ability to find the optima of certain objective functions, subject to some constraints. But they also depend on our willingness and ability to understand how people organize themselves, how they choose objectives and constraints, how they get the relevant data, how they implement decisions, and how they monitor and adjust to the consequences of their decisions. I believe the major future contributions by marketing students to public and private policymakers will come from our willingness and ability to observe and describe organizational decision processes.

Changing our Way of Looking at an Organization

As we all know, certain entities cannot be experienced directly by our five senses. For instance, some entities are too small: no one has ever seen or touched an atomic particle. Some entities cannot be experienced by our senses because they are too big: no one has even seen or touched the economic system of the United States, "Safeway, Inc.," or the "University of California."

There is no best way to represent any entity. For some problems, one conceptualization may be appropriate, but for others, another conceptualization may be preferable. The purpose of this section, then, is to propose one view of an organization that may lead to a useful understanding of organizations as buyers and users of scientific and technical information.

Information, Communication, and Meaning

In his introduction, Joel Goldhar has mentioned that one of the problems we should help resolve is simply "to move information." The challenge posed to us is not in the technical sense of moving signals, symbols, etc. From the criers in the Greek and Roman markets, to the printing press, the telephone and telegraph, the radio and TV media, and now to the coaxial cable and laser, technological development has been "over generous" in allowing us to send an ever-increasing number of signals per unit of time with less noise.

The challenge comes from the fact that our ability to move "meaning" rather than signals has not changed throughout the centuries. Psychology and related disciplines have established at least one basic distinction: physical perception (the perceiving of physical signals) as opposed to psychological perception (the perceiving of the *meaning* of a set of signals). Although the former is relatively well understood, research on the latter has only established that the receiver of signals "sees only what s/he knows"—a fact well known to Goethe and other poets.

The creators of messages and any intervening middlemen want to *communicate,* to assure that the receiver gets the intended meaning. Since the received meaning depends on what the receiver knows, the sender of a message should know what the receiver knows—and here is where the key problem comes in. What a receiver knows, or more precisely, what kind of knowledge the receiver will use to interpret (give meaning to) the message, depends on many factors and conditions, all well documented in laboratory and field work. The high number of possible combinations of factors and conditions makes each application of this knowledge very problematic.

Given this problem of how an individual perceives meaning, let us now consider an individual in an organization. Presumably, each member of an organization makes some decisions. At the very least, each person decides how to receive, use, and send information. More precisely, each member of an organization "processes" information with respect to the following *operations:*

1. identification of a problem;
2. sensing and acquiring from the environment and the organization information that is relevant to the problem at hand;
3. storage of this information in a relevant way (relevant at least with respect to the ability to recall its availability);

4. retrieval of relevant information; and
5. analyses (manipulations) of the stored/retrieved information in a relevant way—that is, reaching a feasible (satisfying) conclusion or decision.

But organizational life does not end here, for the following operations will also occur:

6. implementation of the decision (in itself creating a new set of information);
7. monitoring (feedback) the consequences of the implementation (a new set of information) in a relevant way;
8. and so on, with the unfolding of the dynamics of organizational processes.

A key word in the above list is "relevant." For the moment, it relates to each individual member of the organization—what s/he knows, what s/he expects or hopes, what kind of problem s/he is dealing with, and how s/he perceived it. Information (a bit, a datum, a set of signals) has no meaning in the absolute. Information that is relevant to one person has meaning to that person, but it may have no meaning—or a different meaning—to someone else.

Thus, in an organization, each individual processes (receives, uses, and sends) information, but the meaning of the processed information may vary in degree and kind across individuals. Can we look at an organization so as to understand some of the factors making for variance in the meaning of information?

*Organizational Information Processes
and the Organizational Charts*

I strongly suspect that traditional organizational charts are not a useful way to describe, and thus study and understand, organizational information processes. Organizational charts describe, at best, lines of authority and responsibility. However, the meaning of the terms authority and responsibility is not clear, even in the context of legal cases! In more recent terminology, organizational charts also portray "profit centers." Some prefer to see in an organizational chart a decision tree, with each node in the tree representing a stage of problem solving. This conceptualization is best represented by the ideal Weberian bureaucracy, where the organizational chart depicts a cascade of decision trees, each hierarchically nested in the higher-order tree.

I do not want to argue whether organizational charts do depict some useful aspects of the structure and behavior of organizations. But I submit that organizational charts cannot explicitly capture aspects of organizational processes that are relevant to the buying and using of scientific and technical information.

To begin with, organizational charts are said to describe the "formal" aspects of an organization (perhaps by formal is meant authority and responsibility). Consequently, everything else that goes on in an organization is classified as "informal." For the past several decades, such informal aspects have become the province of human relations and similar approaches. For our purposes, however, this is not an acceptable situation for several reasons.

Since the early work by Bavelas and others to the recent work by Mackenzie, small group dynamics literature suggests that the distinction between formal and informal is not necessary and, in fact, may prejudice the study of organizational information processes. As an illustration, recall the list of operations that each member of an organization may perform—points 1 through 8, above. The acquiring, receiving, using, and sending of information implied by this list will occur in a variety of ways through a variety of channels. If we continue to limit our study of information processed only through formal channels, we may be missing most if not all of what makes an organization tick.

More importantly, and as a generalization of the above criticism of the formal-informal distinction, organizational charts cannot describe the functioning of a group of people. Think of an assembly line, and you will realize that it would not be useful to describe it in an organizational chart format if we wanted to balance it in some optimal way. But a visualization of it by a flow chart—say PERT—does provide a "useful" description.

Thus, for our purposes organizational charts do not identify the factors making for variance in the meaning of information across individuals and groups in an organization. They do not provide a picture of an organization as a processor of information and thus cannot lead us to understand an organization as a buyer and user of STI. Let me suggest another way of thinking of an organization.

Reinterpretating the Inducement-Contribution Theory

In the early 1950s, Herbert Simon proposed the use of the construct of vested interest group as the building block of a new picture of organizations. To illustrate, a business firm comes into being through a cooperative and competitive game among at least four vested interest groups:

1. the suppliers of materials and equipment,
2. the suppliers of personal services (labor),
3. the suppliers of capital (stocks, bonds, etc.), and
4. the suppliers of revenues (the firm's customers).

(I shall overlook the other "publics": the government and the general public. For a discussion of the general public and its impact in deciding about the organizational locus of the public relations function, see Nicosia [1959].)

If any one of these four groups has no members (e.g., if no one wants to work for that firm), the firm cannot exist. To assure the participation of individuals in one or more of the four groups, there must be a long-run balance between the contribution requested by the firm and the inducement it offers. The "game" is cooperative in the sense that the more contributions each participant makes, the higher the output produced; the game is competitive in the sense that each participant would like to increase his/her share (i.e., the received inducement) of the output.

How can a firm assure that the contribution requested and the inducement offered are acceptable to each of the four vested interest groups? By a two-way process of information/communication. This is done by acquiring the information necessary to understand the needs and wants of customers, labor, etc., and by sending information necessary to communicate the advantages of the inducements offered (the product's features, the fringe benefits, etc.).

Not only do the needs and wants of each vested interest group vary in substance, but also the institutional arrangements (e.g., the mass media) that reach each group vary a great deal. Accordingly, the firm structures itself by specializing its information processes by type of vested interest group. Titles of departments—such as personnel, financing, marketing, and purchasing—reflect the need to specialize communications in ways relevant to each of the four groups—the necessary conditions for a firm to exist. These titles may also reflect formal relationships of authority and responsibility, but the raison d'être of such formal relationships is the primary need to specialize the acquisitions and distribution of information by types of people in the environment.

To put it differently, a firm must recognize the heterogeneity in the environment it deals with and specialize its information processing by types of environment to achieve two-way communication with each group. In this sense, information is not a homogeneous good, that is, the same set of data may mean different things not only to labor and Wall Street, but to people in the personnel and finance departments.

Thus, at the core of the problem of designing an organization is the notion of "market segmentation," but as it applies to the entire organizational information process of acquisition of information from different environments, use of information within and across a firm's structure, and sending information to different environments.

To segment the organization of a firm by types of information processes is a question of specialization, i.e., how to achieve better two-way communication with each environment. In terms of organizational design, decentralization is a way to gain the advantages of specialization. In theory, it would be nice to establish communication with each member of the consumer group, with each supplier, etc. Correspondingly, each officer of the organization should receive/acquire information specifically tailored to his/her needs.

But the advantages of specialization come at a cost: the loss of the advantages of coordination. That is, one buys the advantages of specialization in information processing by giving up the advantages of their centralization. In practice, the best organizational design of information processing is to strike a balance between specialization and coordination, between decentralization and centralization of information processing.

An Organization as an Information Processor

Taken together, the points made so far describe an organization as an information processor. I am proposing a view of the structural design (from the board, to divisions, and to departments) and the behaviors of an organization as manifestations of information processing. This processing implies acquiring information from the environment, using this information within and across the internal "units" of the organization, and sending information to the environment.

I have also suggested that the heterogeneity of the environment—the four vested interest groups mentioned earlier—compels a firm to specialize its structure so that each unit can process information with each type of environment. This specialization by environment increases the likelihood of achieving two-way communication with each vested interest group.

However, this concept needs further elaborations. While it may make sense with respect to the acquiring of information from and sending information to different environments, it does not give us insight about the intervening phase—that of "using" information within and across organizational units. In their paper, King and Zaltman refer to this phase with the term "internal communication." This is a most critical area in general and certainly with respect to both the marketing and buying of STI.

Just as the acquiring and sending of information need to be optimized, so does the intervening phase—internal communication. But optimal internal communication—the processing of information within and across organizational units—implies coordination. To illustrate, while reading the financial news one day, a member of the finance department learns of a new patent applied for by a relatively obscure company. The optimization of internal communication in this situation must satisfy at least three necessary conditions: this finance officer (1) must be aware that such a patent may be of interest to some specific officer in the R&D or production departments; (b) must also be aware of how to communicate the information to such an officer; and, of course (c) must be motivated to do so. Unfortunately, the view I have presented does not suggest how such processes of information exchanges occur within and across organization units. The view, in other words, is silent with respect to organizational processes of coordination that may help or hinder the optimization of internal communication.

I am concerned about this shortcoming, for an organization in a modern and technological world is more complex than the simple case of decentralization of information processes by four types of vested interest groups would indicate. From the early work by Ashby in the biological sciences to the recent work in contingency theories on social organizations, we see the development of a view essentially similar to my interpretation of Simon's inducement-contribution theory. In this literature, the structuring of any organism is seen as the general problem of designing an optimal way to process information—to interact optimally with the environment.

Briefly put, an organism receives/seeks information from the environment and attempts to make "sense to its own purposes" of such information. The simpler, more homogeneous the information generated by the environment, the simpler the "organizational" task of the organism to extract useful meaning from the information. As an example, with perfectly homogeneous information, only one reading is necessary. But the less homogeneous this information, the more necessary it is for the organism to design some organizational apparatus to repeatedly sample the available information.

As homogeneity decreases and heterogeneity increases, the more difficult it is for the organism to understand the meaning of the available information. Heterogeneity in information may occur in several dimensions: e.g., reliability and uncertainty. Apparently, organisms cope with heterogeneity by specializing their sending mechanisms.

This view was applied by Nonaka (1972) in studying the

organization of the marketing departments in two industrial and two consumer firms. The expectation was that a direct relationship would be found: high heterogeneity would call for very specialized (decentralized) marketing organization (i.e., grouping by product, and within each product, grouping by brands) and the converse for highly homogeneous environmental information.

While the data seemed to confirm this relationship, many problems came to light (Nonaka and Nicosia [1977]). For instance, one of the firms was a very large electronic company, Hewlett Packard. For one of its major divisions, the prediction was that its product managers would experience high uncertainty in the environment because of the high rate of technological change in the output markets (the potential buyers), in the competitors, in the division's own production processes, and in the division's input markets (the suppliers). But the absorption of this high uncertainty in the division's environment was not reflected by the behavior and structure of the marketing group. It was the R&D department that read and made sense of the uncertainty in the environment.

Thus, the "obvious" idea that the marketing department specializes in dealing with information about the output markets may not be so obvious at all. There must be internal mechanisms that specify the actual tasks of organizational units —a question we become aware of when we look at an organization as an information processor.

Furthermore, there must be internal mechanisms that govern viable processes of communication across organizational units (e.g., communication between the R&D and the marketing departments). The recent literature in industrial procurement seems to raise the same point. On the one hand, the more complicated the input markets—the more heterogeneity in the information they generate—the more specialization we see in the organization of the purchasing department (by product groups, by types of suppliers, etc.). On the other hand, the recent literature and several surveys (see Nicosia and Wind [1977]) show the presence of a complex process of "internal communication" across many departments, one moment of which is the signing of a contract (the so-called "decision," so important to lawyers and traditional researchers).

The point is simple. Previously, I put forward the view of an organization as an organism that processes information via internal specialization by types of environments. Yet, when we look at what an organization does with the acquired information, we have no fundamental framework to give us insight into the processes of internal communication.

Moreover, in reporting a study of marketing departments as processors of environmental information, I am sharing a disturbing fact. As internal communication goes on, we find that information characteristics we thought would raise problems of meaning to marketing departments, do not do so, for the meaning of such information is, in fact, dealt with by some other "departments." (Nonaka's [1972] data even suggest that the perceived "power" of different departments is related not to which official collects what information, but to which official does the interpretation of the meaning of the information.) This fact signals to us the potential complexity of the organization of information processing; that is, it is not clear "who" processes "which" environmental information and "how" this is followed by "which" process of internal communication.

Where Do We Go From Here?

As you can see, so far I have kept my promise of not delivering anything directly useful to the solution of the daily problems faced by the sellers and buyers of STI. Let me next discuss some implications that may be useful for the sellers and buyers of STI and then some notions about what we should do to improve our ability as marketers and students of human behavior to help business people and public policymakers.

Some Implications of the Proposed View

The implications we can derive concern the marketing and marketers of STI, as well as the organizations that may be interested in obtaining STI. Echoing the premise that some of us preached in introducing the notion of consumer decision processes almost twenty years ago, understanding how an organization goes about acquiring, using, and sending information is a premise for successful efforts by the sellers of STI. This understanding is also a premise for designing optimally the structure of an organization's information processing, including the buying and using of STI.

The first implication concerns the question of who collects, stores, uses, or filters STI. A formal and traditional answer would be the R&D group, the engineering group, and/or the production group. But this is too restrictive. Other groups often participate. In addition to the findings by Nonaka, in high-technology firms the purchasing group is frequently held responsible to be on the lookout for information (in fact, salespeople often know that passing information or even gossip on who has discovered or bought what is crucial to a purchasing manager). In some firms, their own salespeople bring back information from, or requests by, the firm's buyers concerning certain technological issues. In corporate planning, especially in cases of acquisitions and diversification, STI is sought and interpreted even by lawyers and finance officers. In a word, it is dangerous to have preconceived notions of which organizational unit collects, stores, uses, or filters STI.

A second implication, somewhat related to the first, concerns the distinction between scientific and technical information. Especially in the context of a modern world exposed to an increasing rate of technological change, the usefulness of the distinction is questionable. At the very least, what is today scientific may well become technical tomorrow (recall the recent case of the laser beam). It would thus be dangerous to organize a firm so that the R&D group is responsible for seeking "scientific" information and the engineering and production group for "technical" information.

The organizational design should actually develop mechanisms of internal communication to facilitate the transfer from scientific to technical "status." (See, e.g., the literature on departmental loyalties, role theory, motivation and cognition, and personality.) Of particular interest are those internal transfer mechanisms where organizational units that have nothing to do directly with STI intervene as mediators.

Consider a few examples of "mediation." The transfer of information regarding the oxygen furnace from a scientific fact to a production reality in some large U.S. firms was filtered heavily (i.e., delayed) by financial-fiscal considerations. The acquisition and use of STI by the railroads was filtered for decades by the interaction between the personnel group and the unions and by the historically unbelievable role of the Interstate Commerce Commission. In the sixties, a firm introduced "error-free" disk packs for computer systems. Mass media advertising was used to convey some of the technical information about these disk packs, and the intended audiences were EDP managers, programmers, and purchasing managers. Unfortunately, machine operators turned out to be more than just filters—they simply blocked the purchase and use of this new type of disk.

Thus, we can see that, at least in principle, any organizational unit can get into the act of acquisition and internal communication of STI. That is, just as the buying of material and equipment is a process very diffused throughout an organization (Nicosia and Wind [1977]), the "buying" of STI may in fact also be very diffused.

I am aware that industrial salespeople long ago discovered the points I have made, and that this awareness is being discovered also by those who choose media in industrial advertising. Publishers of textbooks have not yet caught up with the multitude of buyers/influencers/users.

To the extent that the STI industry is young, it may not yet have experienced the necessity to calibrate its marketing strategies to the complexity of internal communication within an organization. But to the extent to which the STI industry is reaching a point of saturation relative to its current practices, the industry may find it necessary to look more closely to the distinctions among "who does the acquiring, storing, retrieving, and using of STI."

There is another implication of the view I have presented that may one day turn out to be relevant. It is a challenge to some of the key practices by STI marketers—namely, their coding criteria and their strategies in product development and market segmentation.

At the individual level, people differ in personality, cognitive styles, etc. These differences may well affect the meaning given to any information, including STI. Further, department loyalties and professional identification do affect the individual's process of interpreting information. Each individual will tend not only to decode and reincode the received information in his/her own way, but more importantly, s/he will also store (in the mind or in the files) the meaning of the received information in a personalized way. The storing process thus affects the retrieval process.

This variety of individual ways of giving meaning, storing, and retrieving is a formidable organizational problem. For instance, we all know the many troubles met by attempts to organize so-called Management Information Systems (MIS). After all, some evidence—from personal experience to scholarly research—indicates that even personal liking for printout formats affects executives' judgment (the meaning of a set of information bits vis-à-vis a problem, a decision). Another familiar case is the library. Unless the user already knows a lot about the topic of inquiry, s/he will experience the library as the most ingenious creation by a rat psychologist.

Libraries, MIS, and other types of data base are constructed by using some kind of "objective" set of criteria to give meaning to a book and thus classify it. (Accordingly, my book on consumer decision processes was classified by the Library of Congress under the heading of home economics!) Then, funds are spent to apply statistical inference to maximize the amount of information retrieved from the data base that is relevant to the inquiries. But we cannot expect miracles from statistical virtuosity, for the problem is not in the retrieval but in the encoding—that is, in the criteria used in storing a bit of information.

The encoding/storing criteria currently used are based on the assumption of high homogeneity of meanings across users, thus the belief that such criteria are "objective." But I argue that there exists a potentially high variety of meanings across individuals and organizational units. Thus, the encoding/storing criteria should reflect the variations in meaning that different groups of people will use in approaching a data base. If these personal meanings do not sufficiently match the so-called objective criteria, the retrieval cannot be satisfactory. In fact, the

lower the match, the lower is the expected value of the retrieved information to the user.

I understand that the STI industry codes and thus develops
its products by using a few and traditional objective criteria. If
an article deals with electrical engineering, then its potential
users are electrical engineers. During its youthful period, the
STI industry can afford this approximation. But as the distinction becomes blurred between what is electrical amd what is
chemical in some research domain, the chemical engineer is a
lost buyer for the industry and a lost opportunity for his/her
organization.

To put it differently, the strategy used by the STI industry in
its "market segmentation," and thus product development and
marketing, may become inefficient in the near future. To the
extent that meaning changes across organizational units and
people, and to the extent that different people and units participate in interpreting STI, especially new information, the
more useful it will be to segment markets, develop products,
and invent marketing strategies around the criteria that reflect
variation in meanings of information across and within social
groups. The successful marketers of STI will be those whose
encoding criteria reflect the "subjective" criteria of STI users.

Some Things to Do

I began by noting how different ways of conceptualizing a
phenomenon are appropriate for different problems, and proposed that we look at an organization as an information processor. I have also stressed that designing the structure of an
organization's information processing is a key to optimal organizational behavior (decision making).

The current and continuing explosion in new and evolving
scientific and technical information will call for improvements
in the design of an organization's information processing, and in
the product development and marketing by the STI industry.
These improvements will depend on a number of tendencies in
the organization of buyers of STI, which we can forecast on the
basis of the perspective presented here. These major tendencies
will be:

1. The growing amount of new and changing STI will be experienced by organizational buyers as increasing heterogeneity—e.g., less reliability, higher uncertainty, etc.
2. The variance in the meaning of the same set of information across different groups and individuals within an organization will increase.

As a result of these two tendencies, the meaning(s) of information for decision making will become more difficult to establish. The organizational response to this increasing difficulty will unfold thusly :

3. In any organization there will be a tendency for more groups and individuals to be involved in the acquisition, storage, retrieval, and use of STI. This will occur as a result of two organizational necessities:
   3a. On the one hand, heterogeneity in the STI information (i.e., variety in Ashby's view) will require increasing specialization in who processes what information, i.e., decentralization of the organization information processing.
   3b. On the other hand, it will be necessary to counterbalance decentralization's costs by devising coordination mechanisms facilitating communication within the buyers' organizations, i.e., by centralization.

If one is willing to accept this point of view and the tendencies that emanate from it, then practical implications are obvious. To begin with, we can ask: What do we know about these tendencies, theoretically and empirically? Moreover, to answer this question, we would have to have some reasonable knowledge describing the current practices of STI sellers and buyers.

I have the strong impression that we do not even have adequate data concerning these practices. Yet, how can we suggest improvements to the STI industry if the industry itself cannot provide us with the necessary empirical description? After all, our contributions since the sixties to such industries as packaged consumer goods, durables, pharmaceuticals, advertising, and so on have been possible because of the availability of fine-grain data and the willingness of these industries to get such data when not available!

These questions call for many things to be done. But I would suggest that a change must first occur. Both practitioners and researchers should be willing to reexamine the prevailing ways of thinking and feeling about the selling and buying of STI. We should become aware that the obvious may not be obvious upon closer scrutiny, and we should be willing to learn how to question the obvious.

STI is a very heterogeneous good at any one point in time, and especially over time. For the sellers, the current practices of segmenting the potential users and, accordingly, developing and distributing the STI products may become a major hindrance to improvements for everyone concerned. Organizational buyers of STI cannot continue to cope with its heterogeneity by further decentralization in organizational design of their information processing. Beyond a certain level, decentralization rapidly approaches entropy in the sense of lacking organizational order. Some marketing researchers may believe that this has already happened—if it is true that individuals as consumers behave stochastically, then this may also be the case when they behave in groups!

For organizational buyers in particular, the thing to do is in principle clear: search for some dynamic suboptimization between degree of centralization and degree of decentralization in the design of the structure of an organization's information processing. This is a challenge, for we have been busy doing other things (Nonaka and Nicosia [1977]): some have been studying optimal decision rules while others have been concerned with power, human relations, motivation and the like. But practically all of us have neglected to look at the underlying structure of decision making—the operations concerning the processing of information (see points 1 through 8 listed earlier).

What do we know about the substance of all these aspects of organizational behavior? As a yardstick, we may think back to the time when computers began to be bought by organizations. Practitioners and researchers agreed that computers would centralize information processing and, therefore, decision making! It took about ten years of experience to discover that computers were in fact facilitating decentralization, especially in the area of tactical decisions.

This volume is an "unobtrusive" indicator that we have not made much progress in understanding a fundamental aspect of the structure and the behavior of organizations—information processing. That we have little knowledge of how organizations buy STI beyond the experience and insight of practitioners is only a specific symptom of our ignorance of the general case.

Last but not least, there is another thing about which we should start doing something. This concerns public policy. In recent years, as I lifted my sights to broader questions about consumer behavior, I discovered public policy. In the area of advertising and general mass communication, I found that public policy is either nonexistent or ad hoc, inconsistent, and based on ineffable insights, personal values, and lack of facts (Nicosia [1974], Part Two). In the area of technology and consumer behavior, I found it almost embarrassing to report the results of a study of several of the hundreds of federal agencies dealing with consumers (Mayer and Nicosia [1974]).

In a word, I have come to agree with de Sola Pool (1974) and others. He writes: "Technological change transforms entire communications systems. Alternative ways of organizing systems confront societies with difficult choices. Hard knowledge is needed to make the policy decisions of the future" (abstract). I thus welcome the increasing interest in finding ways to improve a component of society's communication system: the exchange of scientific and technical information. My hope is that I have at least successfully argued that we need hard knowledge—i.e., appropriate points of view leading to appropriate constructs and thus, eventually, data that can be used for enlightened action by STI sellers, buyers, and public policymakers.

References

de Sola Pool, Ithiel. "The Rise of Communications Policy Research."

Journal of Communication 24 (1974):2.

Mackenzie, K. D. *A Theory of Group Sructures.* 2 vols. New York: Gordon and Breach Science Publishers, 1976.

March, J. G., and Simon, H. *Organizations.* New York: Wiley, 1958.

Mayer, R., and Nicosia, F. M. "Technology, the Consumer, and Information Flows." Vol. 4. In *Technological Change, Product Proliferation, and Consumer Decision Processes.* Edited by F. M. Nicosia. Institute of Business and Economic Research, University of California, Berkeley. Washington, D.C.: Office of Policy Research and Analysis, National Science Foundation, 1974.

Nicosia, F. M. *Advertising, Management, and Society: A Business Point of View.* New York: McGraw-Hill, 1974.

——. "Public Relations: A Managerial Approach." Paper read at the First Asian-American Public Relations and International Cooperation Conference, Tokyo, Japan, Waseda University, 1959.

Nicosia, F. M., and Wind, Y. "Behavioral Models of Organizational Buying Processes." In *Behavioral Models for Market Analysis: Foundations for Marketing Action.* Edited by F. M. Nicosia and Y. Wind. New York: Holt, Rinehart and Winston, 1977.

Nonaka, I. "Organization and Market: Exploratory Study of Centralization vs. Decentralization." Unpublished Ph.D. dissertation, Graduate School of Business Administration, University of California, Berkeley, 1972.

Nonaka, I., and Nicosia, F. M. "Marketing Management, Its Environment, and Information Processing: A Problem of Organizational Design." *Journal of Business Research,* 1979.

Sheth, J. N. "Recent Development in Organizational Buying Behavior." Working paper, College of commerce and Business Administration, University of Illinois at Urbana-Champaign, 1976.

Webster, F., and Wind, Y. *Organizational Buying Behavior.* Englewood Cliffs. N.J.: Prentice-Hall, 1972.

13
An Agenda for the Future

Joel Goldhar

The preceding papers and the conference discussion amply demonstrate the potential for the application of marketing theory. Clearly, however, there is much to be done before this impact will be felt in a measurable way.

I would like to try to summarize a little bit of what has gone on by identifying what I personally consider to be the areas of greatest potential. I should emphasize that this is my personal judgement based on this conference, the papers and discussions, and on my own previous experience. It does not imply National Science Foundation policy or future research plans. My personal hope is that the STI services and the marketing community will be able to work together to address these problems.

First, it will be important to develop product design guidelines as a function of both the *use* and the *user*. To do this will require a differentiation between the traditional "product" and "service" characteristics of information and the development of a better understanding of the difference between research and engineering users, users in various fields, and user segments within disciplines.

The issue of a marketing strategy for STI organizations must also be addressed. How can product awareness be created? How can products be positioned to take advantage of their design characteristics and their differential appeals to market segments? What product introduction strategy should be used? How will new technologies affect the appropriateness of these strategies?

These broad management issues require that a number of

subissues be addressed. For instance, what is the relationship between "packaging" and distribution channels? What channel choices will most effectively allocate scarce resources? What is the elasticity of demand for information and information services in each use segment?

Many promotion-oriented issues must also be addressed. How can user attitudes be changed to permit greater acceptance of new information? Can effective marketing training programs be developed for the information industry?

Each of these management issues must be based on basic knowledge that needs to be developed. For instance, can "work-style" be identified and used in the way that "life-style" has been in other marketing contexts? What are the roles and relationships of the various actors in the information purchase decision? Which products are substitutes for each other? What are the cross elasticities? What role does STI play in the acquisition of technology? From the government point of view, the public policy issues are also complex. How can the process by which innovations occur be facilitated through information? What is the impact of the regulatory and legal systems on STI transfer? The list of important unanswered questions and research issues is a long one. Yet, its very size and complexity are a challenge to both marketers and STI practitioners. I hope that this conference and this volume will set the stage for addressing these issues with greater specificity.

Participants in Conference on Marketing Scientific and Technical Information

Held at the University of Pittsburgh

S. K. Cabeen
Engineering Societies Library
New York, New York

Alok K. Chakrabarti
Associate Professor of Management
    and Organizational Sciences
Drexel University

Frederick S. Cushing
Director of Publishing Services
Instrument Society of America
Pittsburgh, Pennsylvania

Joel Cohen
Professor
University of Florida
Gainesville, Florida

Anthony Debons
Professor of Information Science
Vice Chairman, Interdisciplinary
    Information Science
University of Pittsburgh
Pittsburgh, Pennsylvania

Rohit Deshpande
Research Assistant
University of Pittsburgh
Pittsburgh, Pennsylvania

Michael Etgar
Professor
State University of New York
    at Buffalo
Buffalo, New York

Robin A. C. Fearn
Coordinatory, Information for
    Campus and Commerce
University of Florida
Gainesville, Florida

Edward Field
Man Labs, Inc.
Cambridge, Massachusetts

Eugene Fram
Professor of Marketing and Director
Center for Management Study
Rochester Institute of Technology
Rochester, New York

Eugene Garfield
President
Institute for Scientific Information
Philadelphia, Pennsylvania

Joel Goldhar
Program Director, User Support
    Program
National Science Foundation
Washington, D.C.

John F. Grashof
Associate Professor
School of Business Administration
Temple University
Philadelphia, Pennsylvania

Ed Howie
Director, Knowledge Availability
   Systems Center
University of Pittsburgh
Pittsburgh, Pennsylvania

William R. King
Professor of Business Administration
University of Pittsburgh
Pittsburgh, Pennsylvania

David Lakamp
Vice President
Smithsonian Science Information
   Exchange, Inc.
Washington, D.C.

Richard Lee
User Support Program
National Science Foundation
Washington, D.C.

John Memmolo
American Institute of Aeronautics
   and Astronautics
New York, New York

Anita Newell
Research Librarian
Westinghouse R&D Center
Pittsburgh, Pennsylvania

Franco Nicosia
Professor
University of California at
   Berkeley
Berkeley, California

Robert Perloff
Professor of Business and Psychology
University of Pittsburgh
Pittsburgh, Pennsylvania

Edgar Pessemier
Professor
Purdue University
West Lafayette, Indiana

Artur Ravin
Principle Co-investigator
Department of Economics
Carnegie Mellon University
Pittsburgh, Pennsylvania

A. Hood Roberts
President
Roberts Information Services, Inc.
Fairfax, Virginia

Martz Robbins
EDUCOM
Princeton, New Jersey

Robert Rothberg
Professor
Rutgers, The State University
   of New Jersey
Newark, New Jersey

Jagdish Sheth
Professor
University of Illinois
Urbana, Illinois

Robert Shriner
Director, Aerospace Research
   Applications Center

Indiana University
Bloomington, Indiana

William Strong
Deputy Director, Center for
    Public Affairs
University of Kentucky
Lexington, Kentucky

Robert Taylor
Dean, School of Information Studies
Syracuse University
Syracuse, New York

Robert Thomas
University of Pennsylvania
Philadelphia, Pennsylvania

Ronald Thornton
Deputy Director, Technology
    Utilization Office
NASA Headquarters
Washington, D.C.

Richard Van Horn
Vice President for Business Affairs
Carnegie Mellon University
Pittsburgh, Pennsylvania

Mary Vasilakis
Research Librarian
Westinghouse Nuclear Center Library
Pittsburgh, Pennsylvania

M. Venkatesan
Professor
University of Iowa
Iowa City, Iowa

Yoram Wind
Professor
University of Pennsylvania
Philadelphia, Pennsylvania

John Wish
Professor
University of Oregon
Eugene, Oregon

Gerald Zaltman
Albert Wesley Frey Professor
  of Marketing
University of Pittsburgh
Pittsburgh, Pennsylvania